The Fatherland Files

The Fatherland Files

Alan N. Clifford

A K Peters
Wellesley, Massachusetts

f/c
(mystery)

Editorial, Sales, and Customer Service Office

A K Peters, Ltd.
289 Linden Street
Wellesley, MA 02181

Library of Congress Cataloging-in-Publication Data

Clifford, Alan N.
    The fatherland files / by Alan N. Clifford.
      p.   cm.
    ISBN 1-56881-034-2
    I. Title.
    PS3553.L436F37   1994
    813′ ,54—dc20   93-45741
                                        CIP

Printed in the United States of America
98 97 96 95 94   10 9 8 7 6 5 4 3 2 1

For J, with thanks to
M, A, and T

Prologue

Kokkina, Cyprus, December 20

Cool, brisk night winds swept across the northern coast of Cyprus. They were not unpleasant breezes, El-Kahrami decided, not after the harsh British bone-chillers. The British weather had been miserable and because of it there was no question in his mind that he had earned his pay in London, putting in long hours and suffering through days in which unending rains froze him to the core. But it had been worth it, he reflected in the comfortable setting of the Topaz restaurant, sipping his third gin and tonic of the evening. The mission had been successful and in retrospect easy, unbelievably easy—though his superiors in the Libyan Army didn't have to know that. As far as they were concerned, it was hard work, and it was only fair that he take a few days off to enjoy himself before flying back to Tripoli. Not that he was going to *tell* them he was taking time off, but just in case they found out, he was ready to defend his actions.

He had been to this resort town on the northwest coast of Cyprus on three other occasions during return trips from Europe and he found it offered several advantages over the more popular resorts on the southern coast. Most importantly, it was off the beaten track. He could drink and carouse as he wanted; no one was likely to notice. And, if anyone did notice, he was just another pleasure-seeking "Egyptian businessman," according to his identification papers. In that guise, the chances were slim that he would be of interest to either Israeli or NATO intelligence agencies, his two main sources of concern.

The whole island—or at least the Greek-speaking part—was getting ready for Christmas, and even though El-Kahrami was not a Christian (nor much of a devout Muslim, for that matter) he enjoyed the colorful lights and friendly smiles that the holiday season brought. This day had been spent leisurely wandering through the fishing areas in a heavy sweater, planning his night's adventure. The Topaz was by now a favorite of his. The few tourists in Kokkina at this time of year rarely ventured through its doors and, best of all, the proprietor could be relied upon to supply a different woman for every night of his stay. Tonight's was a tall blond from Stockholm. She nestled easily in his arms, sipping wine while a strolling guitar player made his rounds.

"Not long. The liquor relaxes me. This way, I'll fall asleep after an hour or so in bed with you. Then you can leave and make some more money with someone else. It's a better deal for you." He laughed and she laughed. It was true; he was a very generous man. After all, what did he care if she only stayed for a short time? Tomorrow there would be another new face and new body to play with. "Right now, you must excuse me. I'm off to the w.c. Too much to drink, time to answer the call of nature." El-Kahrami was 50 and already his prostate was bothering him. He was afraid that soon he'd be like his father and have to lug around a jar with him for emergencies.

The men's room was on the far side of the restaurant and he moved quickly toward it through a maze of half-filled tables. The urinals were a welcome sight. With a sigh of relief, he emptied his bladder, re-zippered his fly, and stopped at the sink to wash his hands. The restroom seemed empty, but before he had a chance to reach for a towel, he felt something hard jam him in the small of his back. He knew immediately that it was the nozzle of a pistol and that someone unfriendly had found him, despite his precautions.

He looked into the mirror above the sink. There were two men reflected in it, one on each side. He couldn't see their faces clearly, but one had blond hair. Probably Americans.

"What do you want?" he asked warily in English.

"A few minutes of your time, Major El-Kahrami," the blond voice answered in Arabic. "A few minutes, that's all. None of your colleagues know you're here, you're all alone, so don't be a hero."

The Libyan's face sagged. Americans didn't bother to learn Arabic. They must be Israelis. Just what he didn't need. Beads of sweat appeared on his forehead.

"Follow my friend out of the restaurant," the voice said. "I'll be behind you with a pistol half way up your ass. Don't do anything foolish. Remember, we're all professionals and no one need get hurt."

El-Kahrami reluctantly followed their orders. They're just on a fishing expedition, he convinced himself, they don't know anything. Despite that consoling thought, the sweat began to run down his neck. A holiday in Cyprus no longer seemed such a clever idea after all.

PART 1

Frankfurt 368

Frankfurt International Airport, January 3

Light snow was falling as TWA flight 412 landed in the fading light of a wintery afternoon. Surrounded by one hundred ninety-eight other weary passengers, Goodman made his way slowly through the terminal building. Although free of its everyday bustle and noise on this quiet Sunday, the Frankfurt terminal contained a full allotment of security policemen.

The reason for the heavy security force was well known to the passengers. On the preceding day, a terrorist attack had taken place in the same area that they were walking through. Grim reminders were still very much in evidence: blood-stained carpeting, bullet holes in the walls and ceiling, shattered windows. Two innocent people dead, seven wounded,

1

two unrepentant assailants—reportedly Arabs—under heavy guard in a Frankfurt police station. Goodman stopped and stared at the blood stains. The sight of dried blood was nothing new to him, but seeing it in massive amounts in an airport . . .

Perhaps because of Goodman's rumpled, unshaven appearance, two of the security police hovered nearby. Tall and muscular, with hard-set faces and submachine guns cradled menacingly across their leather jackets, they scrutinized him carefully.

"Not a pretty sight, is it?" one of them said in halting English.

"Pretty sickening," Goodman replied in German.

At the sound of his native tongue, the policeman grunted in surprise. "You're not a tourist?" he asked, this time in German.

"No."

"May I see your passport, please?"

Goodman handed over his passport. "I'm Dr. Paul Goodman. I'm here to attend a scientific research meeting," he explained.

The policeman studied the passport carefully. Satisfied that Goodman was no threat to the peace and tranquility of the Federal Republic, he returned it with a curt nod. Goodman continued down the corridor.

The line at customs was mercifully short. Coming out of the terminal building, Goodman put up the collar of his overcoat. The snow showed no signs of abating. At the head of the taxi line, a brand-new black Mercedes sedan was waiting, the driver deep into the sports pages of the Allgemeine Zeitung. Goodman threw his valise and attache case into the rear seat and slipped in next to them.

"Rotten weather," he said disgustedly.

"Could be worse," the driver replied. "Where to?"

"Downtown. Hotel Fürstenwald."

The driver put away the paper and switched on the meter. Goodman settled back against the rear seat.

The driver said little, preferring to chain smoke one cigarette after the next. Goodman enjoyed the silence. It gave him the opportunity to watch the passing countryside without distraction. Soon the repetition of industrial parks interspersed with clumps of denuded trees grew boring and he focused instead on the swirling clouds of white flakes battering endlessly against the window. Snow fascinated him, conjuring up memories of childhood in Rochester, New York— sleds, brick fireplaces, hot cider, grandmother's stew, and the interminable winter storms whistling across the Great Lakes. So long ago, he thought, remembering his parents and grandparents staring with red-cheeked pride at his lopsided snowmen. They were all dead now, but at least they had lived long enough to see him graduate from medical school. Their dream and his.

Cigarette smoke drifted into the back of the taxi and he opened his window a crack. Despite the snow, the driver had floored the accelerator and the blast of icy air from the open window stung Goodman's cheeks. But it was not an unbearable sensation, and he kept the window open. Better that than the smoke.

The snow hardly made an impression on the concrete roadway of the autobahn, but it was a different story on the suburban streets that were now on either side of the highway. There the granular flakes covered the tiny front yards and roofs and gables of the small houses arranged neatly in rows. Behind them, the skyscrapers of Frankfurt loomed in the grey mist. Because of the light traffic, the trip was fast. Too fast. Only when they exited from the autobahn in the downtown area was Goodman convinced the driver finally took his foot off the gas pedal.

The hotel that had been selected for him by his hosts was not one of the newer, gaudier edifices that had sprung up in Frankfurt after the war, but rather an older, small establishment that had been built in the reign of the last Kaiser. Overstuffed armchairs and thick pile carpets gave the lobby an aura of old world comfort, though bare patches in the carpet were visible near the front desk.

"Dr. Goodman from New York," he told the desk clerk, a slightly built, balding fellow who seemed extremely relieved to see him.

"We thought you weren't coming," the clerk said. "Your company told us you would arrive before noon." Then his tone grew more curt. "Our guests often let us know if they expect to be arriving later than scheduled."

"I would have liked to," Goodman replied evenly, "but I didn't know my plane would be delayed three hours taking off and then have to make an unscheduled stop in London for engine repairs. That took another four hours. Then there was a bomb scare. That delayed us another hour. So you see, I did plan to arrive by noon. Because the trip was very long and I'm tired, I don't appreciate your rudeness." He glowered at the clerk, who quickly changed his tone.

"Excuse me, Herr Doctor. It's just that your firm had reserved an excellent suite, 15F, one of our best. We were concerned that we wouldn't be able to rent it if you didn't come." He wiped moisture from his forehead and smiled weakly. "May I say that it is a delight to meet an American who chooses to converse in German." He beckoned to Goodman to sign the

4

ledger. "A little unusual to have snow this early in January," he added apologetically, "but then it will soon melt, that I am sure of."

"Good," said Goodman, returning the ledger, "I want to see a little of your city before I leave."

"Yes, you must," the clerk agreed, warming to the subject, "perhaps even today. Start with a nice walk before dinner. The streets around here are quite interesting. The Liebrauenkirche is very close, with lovely fountains in its square. A wonderful sight with a little snow on it."

"I'm too tired for even that."

The clerk persisted. "A walk in this weather is invigorating. Go see our Römer, the city hall. It's a fantastic old building, well preserved."

Goodman shook his head. "Too tired; I hardly slept on the plane."

"Of course. If you don't wish to take a walk, may I suggest one of our hearty dinners, some excellent beer, and a good night's sleep. Our hotel has a reputation for taking good care of its guests. We have a fine dining room; shall I make a reservation for you?"

"I'll probably order room service. I don't feel like dressing for the dining room."

"Ours is most informal."

"No, thank you," Goodman said forcefully. He had had enough of this. "My room key, please."

The clerk wiped his forehead again and rang for a bellboy.

The suite that was reserved for him had a sitting room and a large bedroom with a separate dressing area and bathroom. Werner's company was doing things in style, he had to admit, but then after that awful flight he deserved it. He examined his face carefully in the bathroom mirror, rubbing the day-and-a-half's growth of stubble. Too much gray in it, especially for someone just turned thirty-five. He had long ago concluded that his face was totally undistinguished, eyes somewhere between gray and blue, nose straight but a bit broad, mouth small and much too severe. Undressing for the shower, he paused for a moment to stare with satisfaction at the reflection of his lean body and his muscular legs, firmed by long hours of running and bicycling in Central Park. He lingered in the shower for what was for him an unduly long time, letting the hot water run over his neck and back to help soothe away the discomfort of his journey. Quickly drying his face and body, he dressed in his most comfortable traveling pajamas, found a news program on the television set, and propped himself in the bed.

Catastrophes blared out at him in rapid-fire succession: more details of Saturday's terrorist bombing at the airport—they were Arabs after all—fierce gunfights in Bosnia, riots in the former Soviet Union, and on and on. The news stories that he had left behind in New York had followed him to Frankfurt and he was sure would be waiting for him in New York when he returned. Too depressing. When he felt the weariness of the day's journey settle over him again, he decided it was time to order dinner. "Venison and cabbage and beer," he told the room service operator. Then he would sleep the whole night through, so that in the morning he would be free of jet lag and ready for the meeting, the meeting that was the whole purpose of this odyssey.

Despite Werner's good intentions, and the hospitality of the company, Goodman was still skeptical about the meeting. He knew Eriksson would be there from Copenhagen. Their relationship was a friendly one and Eriksson often stopped by when he was in New York to exchange data and ideas. But this fellow

Stern—the one who had done some of the first clinical trials with the new drug in Hannover—he didn't know him at all. Not that it really mattered. After all, it was Peter Werner that he was most anxious to see.

He remembered how he had first met Werner at an international cardiology symposium in New York the summer before. As the faculty representative from one of the host medical centers, Goodman was the chairman of a panel session that was discussing the various indications and contradictions for using different drugs to treat irregularities of heart rhythm. After the formal presentations, he opened the discussion to questions and comments from the floor. To his dismay, a questioner from the audience, a tall, good-looking blond fellow with slightly tinted glasses, monopolized the entire question-and-answer period. Probing and incisive in his comments, he pushed the conservative panelists, and Goodman in particular, to an even more defensive posture. The questioner identified himself as Peter Werner. He was obviously European from his accent, and he knew the subject too well from Goodman's point of view. Much to Goodman's chagrin, Werner clearly bested him in what turned out to be an informal debate before the amused audience.

Afterwards, Werner sought him out to make amends for any embarrassment he might have caused. Over coffee, they continued their discussion in a mixture of English and German, but instead of arguing, they were surprised to find themselves in agreement on most aspects of the problem.

"You see," Goodman said with a triumphant smile, "I'm not really as much of an ogre as you thought."

"No," Werner admitted, "I was wrong about that, though you do tend to be a trifle reluctant in these matters, especially when it comes to newer technology."

"My first rule is do no harm."

Werner was nonplussed. "Yes, but if these new approaches can help patients, we must find ways of using them."

"You say *we*, but I see from your name tag you're with a pharmaceutical firm. Were you in practice at one time?"

Werner sighed. "I wish I had been, but the answer is no. My father was also a physician, the director of research for my present company, and he wanted me to follow in his footsteps. So after medical school, I joined the firm. Needless to say, my father was a very strong-willed person. After he died I took his place as research director. I enjoy it. We're a small firm, Frankfurt Pharmaceutica. We're quite active in Germany, but not that well known internationally. I have a free hand in research and development and, as you gathered, cardiovascular pharmacology is my major interest."

"From your questions this afternoon," Goodman said, "that indeed was what I gathered."

Werner laughed, and to his own surprise, Goodman laughed too.

From that encounter, the beginning buds of a friendship emerged despite the obvious contrasts between the two men: Peter Werner, the blond, German scientist with his precisely accented English, and Paul Goodman, the dark, curly-haired New York Jew with a tendency to speak too quickly and too nasally. They met several times in the next few days, attending sessions together, listening to the presentation of papers, discussing the implications with other interested colleagues. After the symposium was over, Goodman invited Werner to join him for dinner. Still later, they stopped for a nightcap in Goodman's Upper West Side apartment.

8

Stretched out on the sofa in front of Goodman's living room window with its panoramic view of the Hudson River and New Jersey shoreline, they sipped cognac while Artie Shaw led his band through "Begin the Beguine."

"I can't believe you're really into big band swing," Goodman said with delight. "I thought I was one of the few surviving remnants of that era. Or is remnant the wrong word, since I wasn't even born then?"

Werner laughed. "For you, it must have been through osmosis into the womb. For me it was different. I learned to enjoy swing during the occupation of Frankfurt by your GIs and later the Voice of America was always playing it. Now, of course, no one does. I shouldn't say no one. There's actually a nightclub in Frankfurt where they play swing one night a week. Half the vocals are in English, half in German. It's really quite a popular place. And not just among middle-aged American tourists, but Germans as well. Next time you're in Frankfurt, let me know and I'll show it to you."

"I've never been to Germany."

Werner didn't seem surprised. "You're Jewish?"

"Yes."

"Then I can understand it," he said simply, looking Goodman directly in the eye. "But we're a different people now, Paul. It may sound trite, but most of us weren't even born, or at most were just kids, when the war ended. The Hitler era is gone, dead forever. Nineteen forty-five changed everything. We will never allow the new united Germany to be anything like its predecessors."

Goodman put down his drink. "I don't dispute that, still . . . "

"Did you lose relatives?"

"No, it's not that. In fact, I don't know many people who did, and, for that matter, I'm not very religious. It's just been easier not to go, or, put another way, there's never been a reason to go." He went to the window. The George Washington Bridge was less than a mile away, all aglow like some enormous electric necklace. Beneath him, cars crept slowly home on the crowded West Side Highway.

"I'll make it easy for you," Werner said, still sprawled on the sofa. "I'll invite you over, all expenses paid."

"How can you do that?"

"My company has a new drug for arrhythmias. You see, anti-arrhythmics are one of my pet projects. We're just starting to do clinical tests and we need expert consultants. We've already had Hawthorne come over from London. Now we need an American. It gives the top brass something to brag about."

"What kind of drug is it?"

"It's similar to quinidine, but only needs to be taken once a day."

"Patients would love that."

"Exactly. We're about to review some clinical trials that Stern did for us in Hannover, plus some animal studies. Eriksson from Copenhagen is interested. Why not join us? I can set it up just after the New Year. You'll see my home, my family, and my fellow Germans. Maybe even do some skiing. You speak the language well enough to make it a really worthwhile trip."

"I'm not sure—"

"Why not? Think about it while I hear some of that 1940 Glenn Miller album you showed me before."

For the remainder of the evening, they did not discuss Werner's offer. But several weeks after he left,

10

a formal invitation from the president of the company arrived, along with some information about the drug from Werner, now writing in his capacity as director of research. Werner made it clear that if Goodman liked what he saw and heard at the Frankfurt meeting, the drug company would be pleased to fund any clinical trials that he would want to conduct in the United States. Goodman wrote back his acceptance, found someone to see his patients in his absence, and arranged the trip to Frankfurt, and here he was, for better or worse.

The shower had helped him forget the unpleasant flight and the delay in London, and he was in a much better mood than when he had registered at the hotel. He let the room service waiter in at the first knock, watching with hungry anticipation as he set the table. Two great steins of dark beer accompanied the meal.

"This is with the compliments of the desk clerk, Herr Doctor. He sends it with deepest apologies for his rudeness earlier."

Pleasantly surprised at the gesture, Goodman tasted the beer. It was dark and cold, with a deep body. "Thank the clerk for me."

He ate and drank ravenously. The venison was fine, the cabbage delicious, but the beer had a metallic taste after the first swallow. It was a good try on the part of the clerk, he reflected, as he finished the second stein, but he would stick with lighter beer. Placing the tray outside the door, he left word with the hotel operator to wake him at seven the next morning. He switched off the television set, arranged his wallet and watch on the dresser in the bedroom, leafed idly through the confidential drug dossier that the company had provided him and placed it beside the wallet, then drew the curtains on the still snowy evening. His nervous energy was rapidly dissipating, and he was feeling the effects of the long flight. He also thought he was somewhat dizzy from the beer. When his head met the pillow, he fell asleep almost immediately.

By nature, Goodman was not a heavy sleeper. Perhaps this was due to the incessant din of Manhattan's automobile horns and fire and police sirens, or perhaps it was due to the early morning cacophony of garbage cans being hurled onto sidewalks and trees by sadistic sanitation men. Whatever the reason—or combination of reasons—New York had conditioned him to wake easily. This conditioning carried over to other situations as well, and even in the relatively quiet setting of an upper floor of a not very crowded hotel with a minimum of traffic outside, he expected to toss and turn with regularity. But tonight he slept more soundly than he usually did.

The creak of a floorboard woke Goodman in the middle of the night. His eyes snapped open. The room was dark, but there was a strange clicking sound coming from the dresser. He turned on his side and the clicking stopped. Usually, Goodman would be wide awake in a few seconds when his sleep was interrupted, but tonight his head felt heavy, sodden, his eyelids like lead. He yawned, turned over again and was ready to return to slumber when a shaft of light from the hallway fell across his room. The light penetrated Goodman's torpor and he sat bolt upright. "Somebody there?" he mumbled, turning on his bedside lamp. The room was empty. Not knowing whether he had imagined the shaft of light, he glanced quickly at the dresser where his wallet and watch were. They were still there. Confused, he sank back into the bed. His head felt like it weighed a ton. He switched off the lamp again and within a few minutes he had drifted back into sleep.

The wake-up call came exactly at seven A.M., as he had requested. Still groggy, he had to drag himself to the shower. It was only after the water revived him that he remembered his "dream" of the night before. Wrapped only in the bath towel, he carefully checked his wallet. Nothing was gone. The watch was still there. As an afterthought, he picked up the drug company dossier. Some of the pages were bent back in ways that he could not remember doing. Puzzled, he smoothed out the pages and carefully placed the dossier in his briefcase.

Frankfurt, January 4

The headquarters of Frankfurt Pharmaceutica was on the western outskirts of the city. Monday morning traffic—cars, motorcycles, bicycles—was extremely heavy, and the taxi ride from the downtown area was much more arduous for Goodman's throbbing head than had been the arrival from the airport the day before. Even though the traffic was going mostly in the opposite direction, there was enough volume to tie up several of the downtown intersections. The driver swore repeatedly as he was alternately cut off or did the cutting himself. The sudden stops and startings did not help Goodman's headache, a hangover he attributed to the strong beer the night before.

Very little of the light Sunday snowfall remained in the downtown area. But as they drove further out, a thin veil of white lay over what in warmer months must have been grassy areas. The sun was already peeking through early morning fog and it would not be long before all the snow was gone. Goodman rested his head on the back of the seat and closed his eyes.

Despite the traffic, they arrived some fifteen minutes early. The driver circled the industrial park until he found the company sign, a low-key affair with small black lettering. The headquarters building was striking, low and long and of starkly modern design, yet blending in easily with the trees and grass. Although it was five stories high, the adjoining factory had also been carefully designed to enhance the surroundings rather than to intrude. Even the taxi driver was moved by the plant's appearance. "Nice," he said as he counted out Goodman's change. "I wouldn't mind working around here."

14

When the receptionist showed him into the conference room, Goodman was pleasantly surprised again. The room itself was conventional enough: three tiers of U-shaped writing surfaces with padded chairs placed about four feet apart, and at the center of the U a slide projector and screen. But what made the room breathtaking were the rear and side walls. They were entirely glass and offered an uninterrupted view of an immense open expanse of fields, trees and a small pond, all coated lightly with the remains of yesterday's snow. An idyllic spot, he mused, and like the taxi driver, he too wouldn't have minded working in such a place.

"Dr. Werner will be with you in a few minutes," the receptionist said. "May I get you a cup of coffee?"

"Thank you."

He took two aspirins, sipped the coffee, and enjoyed the scenic backyard. Werner entered the room and approached him with a warm smile.

"Paul, so nice of you to accept our invitation. Welcome to Frankfurt. I wasn't certain you would make the trip until I actually saw you in this room."

Goodman rose to shake his hand. "Peter, you proved too persuasive to say no to."

Werner laughed. "Please continue with your coffee. I'll join you myself in a few moments. I must apologize for our weather. Not very hospitable of us to have a snowfall, but I suppose it's no different in New York at this time of year?"

"No different at all. But what's there to apologize about? I just wish there were more of it so I could go skiing with you."

"Ah, you remembered that. Good." His face grew animated. "Our whole family skis, our two children as well as my wife, and they're only ten and twelve. Don't worry, there's enough snow in the mountains. If only you had the time . . . But that's for later. What about your accommodations? Were they comfortable?"

"Quite."

"Excellent. Our firm likes that hotel very much, one of the few older buildings that survived the war. We put up all our special guests there. So much more charming than the new hotels; God, how I hate them." But he laughed as he said it. It was hard to picture Peter Werner hating anything.

"They're easy to dislike. We have our share of gaudy new hotels in New York."

"I know. Remember that awful place I stayed at?"

"I do."

"I hope you enjoy your visit here," Werner said in a more serious voice. "I don't mean just for the sake of our new drug. I would like you to see a little of the non-scientific aspects of my life. To see my—our—human side, so to speak." He seemed embarrassed by his words, yet obviously felt there was a need to express them. "I don't suppose I'm saying this very well, but I hope that when our meetings are over tomorrow, you will have dinner at my home? I would like you to meet my wife and children. If that is agreeable to you, of course."

"I would be delighted."

"Good, good. I was afraid you had forgotten our little chat in New York."

"No, Peter. I remember it quite well."

The door to the conference room opened and the other invited guest, Lars Eriksson, was ushered in.

"You two are old friends, I think," said Werner as Goodman and Eriksson shook hands.

"Friendly rivals," the Dane responded, winking. He was a thin, almost reed-like man, nearly bald, with high cheekbones and an extremely pale complexion.

16

Eriksson helped himself to some coffee and pastry from a tray that the receptionist had wheeled in. On the stroke of nine a stream of half a dozen young men entered the room, speaking German excitedly but in subdued voices.

"My assistants," Werner said by way of explanation. He pointed to several by name and they nodded at the guests in return. "All are quite bright, of course, and I owe a lot to them."

"Where would any of us be without research assistants?" Eriksson wondered aloud and Werner laughed.

The last figure to enter the room was a much older man—in his late sixties or early seventies with ruddy face and snow-white hair. He walked briskly, cutting an imposing appearance with well-appointed three piece blue suit. He put his arm around Werner and patted his shoulder. "Good morning, Peter. I'm looking forward to an excellent presentation."

"We won't disappoint you, Franz. Let me introduce you to our distinguished guests, Doctor Goodman from New York and Professor Eriksson from Copenhagen. Gentlemen, this is Dr. Franz Richter, the general director of Frankfurt Pharmaceutica." Handshakes were again exchanged. The young assistants, taking their cue from Richter's arrival, quickly took seats in the second and third tier of seats.

"Where is Doctor Stern?" Richter asked.

"He called to say the morning flight from Hannover was delayed because of fog," Werner said. "He should be in by noon, but we can begin without him. The morning will be confined to basic research and the animal studies and he is very familiar with that aspect already." The receptionist handed him some papers. "Ah, excellent. Here is the schedule of activities for the session. Even allowing for a last-minute

17

change or two, I think we can still stick pretty much to the allotted time."

"You've even accounted for Stern's tardiness," Richter said in admiration as he examined the list. "You have him on for the afternoon."

Werner smiled. "That was pure luck."

"Nonsense, Peter, it was good planning. It's characteristic of you."

"Just luck, Franz." Goodman wondered if there was a trace of sarcasm in Werner's reply. The receptionist distributed schedules as they took their seats. Goodman's headache had improved considerably and he was looking forward to the presentations.

A portable lectern had been set up near the screen and it was from this central location that Richter made his opening remarks. Grasping the lectern firmly in his powerful hands, and speaking in robust tones, he welcomed both Goodman and Eriksson as eminent cardiologists invited to share in the results of Frankfurt Pharmaceutica's experimental work. "We earnestly hope that soon Frankfurt 368, this new and still unnamed drug of ours, will be able to be used in patient trials in the United States and Denmark. Dr. Peter Werner, our chief of research and development, will moderate today's session." Turning the podium over to Werner, Richter sat down in the front tier of seats.

Werner rose and with his hand quickly brushed his hair from his forehead. Speaking softly from the lectern, he described the background of the meeting. "In the course of working with some of the standard antiarrhythmic agents, it became apparent to some of us that the results were not very impressive in terms of sustained action. Therefore, we began to look at a new class of agents. The one that we eventually focused our main attention on, 368, was actually discovered by my

18

father. As many of you know, he had been doing extensive work in this area for some time, but had been dismayed by the toxic side effects of the intravenous injections. He felt strongly that it was the adjunctive compound rather than the active ingredient that was responsible for the toxicity, but he never was able to develop another satisfactory adjunctive before he died. Through the efforts of my chief technician, Dieter Hoffman," and here he gestured in the direction of a lanky young man seated in the last row, "a less toxic adjunctive was developed. We were able to work with the new compound in an isolated rabbit heart and finally in an intact dog preparation. The results are impressive and I'll let Dieter describe them for you."

Curtains were drawn on all sides of the room to allow slides to be shown. Hoffman's presentation involved diagrams, charts and electrocardiographic tracings. When it ended, a short question-and-answer period followed. Eriksson had been busily taking notes; now he stopped and leaned toward Goodman. His normally poker face was more alive. "A nice bit of work. Werner's got a good crew, no question about it."

"What they've presented so far has been impressive," Goodman said, "but I'm concerned about the two aspects they haven't gone into yet—the immediate and late toxic effects."

Eriksson nodded. "Yes, of course, those are the potential stumbling blocks. Still, I have a feeling that these chaps have the answers. I don't think they would have hauled us all this far—and I'm thinking specifically of you—if they weren't sure of what they were doing."

As it turned out, Eriksson was correct. In the remainder of the morning session, these very points were discussed by Werner's other assistants. Data was presented that showed no immediate side effects of the drugs in any of the animals studied, nor were there deleterious effects noted up to six months after the last dose had been given.

With the last presentation of the morning now concluded, the curtains were drawn back and bright sunlight filled the room. With a great deal of stretching, the group adjourned to the company cafeteria that was on the same floor as the conference room, but further to the side. A small table had been set aside for Richter, Werner, Goodman, Eriksson and the missing Stern. Luncheon was an assortment of cold meats and vegetables. Beer was available as well, but Goodman decided he hadn't fully recovered from last night's brew. The assistants and technicians sat at another, longer table at the rear of the room, which was now filling with the rest of the company's employees. There were about 100 people in all in the room and Goodman noticed that except for Richter and a scattering of others, the vast majority were quite young. One wall of the cafeteria was entirely glass and looked out at the same open space visible from the conference room. Goodman commented on the welcome change from the dreariness of most of the lunchrooms he had been accustomed to.

"Yes, we're very pleased with the way this building has turned out," Richter said, carefully picking some fat off a slice of roast beef. "We're not a large firm, but our work record has been good, and our products have been well received. We pay the workers well and we give them good surroundings. What more can they want?"

"Are they unionized?"

"Oh, yes," he laughed. "But we have no problems. In this country, the unions and management seem to get along well enough. Not like some other countries." He laughed again and continued picking at his roast beef.

"It's interesting that you should appreciate how pleasant our surroundings are," Werner said to Goodman, "because we take it so much for granted. But it didn't have to work out this way. It was my father who insisted on locating the plant out here in the country. Wasn't that so, Franz? Didn't father actually pick out this spot?"

Richter's face wrinkled. "Well, yes, in a way, I suppose, but his idea was to get out of the downtown area rather than to pick one place over any other."

"It's funny," Werner continued, "but as a boy I remember coming out here with him, even before the plant was built. He seemed to have a great fondness for this place in particular, because I remember the pond so well."

"No," Richter insisted, his voice rising a bit and his knife and fork now working vigorously at the slice of roast beef. "There *was*—and *is*—nothing special about this place. Just to get out of the downtown area, which was all bombed out and depressing."

"That may be, but . . . "

"I have a better memory of those days, Peter, listen to what I say." The tone in the older man's voice was now much more direct. Werner abruptly dropped the line of conversation entirely and began to eat his lunch. There was obvious tension between the two men, and it was with some relief to Goodman and Eriksson that the fifth guest now made his appearance.

Through the cafeteria door came Stern, raincoat in hand, sweat pouring down a rotund face despite the winter chill outside.

"Ah, Walter," Richter said gleefully. "At last the famous Doctor Stern has arrived." He slapped the late arrival on the back and made a great show of helping him to his seat. Stern appeared to be a younger version of Richter—ruddy, stocky and with only a touch of

21

white in his sideburns. Richter introduced Stern to Goodman and Eriksson as the only man who could sweat in the winter. Richter roared as Stern wiped his face and hands.

"The damn plane wouldn't take off," Stern sputtered as he sat down and immediately began attacking the plate set before him. "It was the fog. Always the fog. I should have taken the train." His cheeks filled up as he wolfed down the food and followed it with a pitcher of beer. "Please excuse my being late, Franz," he said between mouthfuls, "and as for Peter, I'm sure he had contingency plans all worked out."

"You see, I told you so." Richter was now in a more relaxed mood. "I told you Peter had planned for everything, even Walter's late arrival."

The others laughed and even Werner visibly relaxed with Richter's return to good humor.

On their way back to the conference room after lunch, Richter drew Werner aside and whispered into the younger man's ear. When Werner blushed, Richter clapped him on the back and with a big smile walked with him back to their seats.

"Looks like they've made up," Eriksson said to Goodman, "like two naughty schoolboys after a little tussle in the yard."

The afternoon's session was devoted entirely to the effects of the drugs on humans, based on Stern's experiences in Hannover. Once at the lectern, Stern shed his jolly fat man role and was all seriousness. In a well-organized manner, he demonstrated how the drug had had no toxic side effects, first on healthy young volunteers—usually students working for extra money—and then on three patients with varieties of heart disease. He showed electrocardiograms with marked irregularities. After the drug had been given, the frequency was obviously reduced, though still present.

"I'm surprised you didn't even get a further reduction," Werner said, "based on the animal studies."

"That's impressive enough," Stern replied.

"Forgetting that you're in a room of the company that's sponsoring this drug," Eriksson asked, "what do you make of this? I mean, isn't it really anecdotal? Just three patients, you know."

"I agree with you completely," Stern said. "It's much too early to say anything definitive. We can only speculate."

"Even so," Eriksson said, "I'm impressed by the prospects for this drug. The type of disease these people have is often lethal. A long-acting preparation that's truly effective would be an important advance."

A murmur of excitement swept the room at Eriksson's remarks. It was at odds with his reputation as a skeptic, not usually given to such enthusiasm.

"And what about you Paul? What do you think?"

Goodman chose his response to Eriksson carefully. "I share both your concern and your excitement, Lars. It's obvious we need more clinical trials, but the potential is enormous."

"Caution, gentlemen, that is the key." Richter was speaking. "We must be careful in not rushing to conclusions before we know the full ramifications of this drug."

Werner shook his head. "I must disagree, Franz. We should push on with this, get as many research centers involved as possible. It's so important."

Richter held up his hand. "At tomorrow's session we will map strategy. I would like to thank today's speakers. Tomorrow's meeting will be at the same time." He rose, and as if by signal, Werner and his chief assistant, Dieter Hoffman, also rose. Their younger colleagues quickly followed suit. Addressing his two distinguished visitors, the director continued, "Tonight you will be my guests for dinner. Taxis will pick you up at seven-thirty sharp. But first, we would like you to take a tour of our plant. Dr. Werner will guide you. See you tonight."

"I'm afraid we must start with the production section of the plant," Werner said after Richter had left. "It's actually rather boring, but Dr. Richter is very proud of our automated production line and insists all guests see it, so please bear with me. Then we can go to the laboratory area, which you should find more interesting." Werner and Hoffman led the two cardiologists into the main hall of the plant. Goodman found Werner's warning prophetic. The tour of the noisy production facilities was of no interest to him at all.

Eriksson complained frequently, leaving no doubt about his feelings. "We have factories in Denmark, too, you know," he said sharply after nearly an hour of peering down long aisles of white-coated workers packaging a variety of chemical items. "One's just as dull as another."

"Can't you show us anything more interesting?" Eriksson persisted. "What's that chap doing over there, for example, guarding your gold supply?" He pointed to a door at one side of the production line that had a security guard seated in front of it.

"That's an area where more volatile chemicals are prepared," Werner replied. "Because of the dangers of explosives and so forth, Dr. Richter insists that only a small number of specially trained workers have access to it. Even I don't go inside, but Hoffman's been in there a few times. Isn't that right, Dieter?"

Hoffman squirmed uncomfortably. "I really can't talk about it" he said. "Dr. Richter is quite strict about these things."

24

At that moment the door to the guarded room opened and three white-coated young men exited.

"What a grumpy-looking lot," Eriksson said. "They've probably got the last shift for lunch."

Apparently deciding that his guests had had enough, Werner mercifully cut short his tour and with Hoffman at his side, took Eriksson and Goodman to the laboratory area.

He proudly showed them the extensive facilities in his own quiet lair. "The equipment is first-rate, as are the assistants, especially Hoffman here, so I have no complaints."

"But surely Richter must be a difficult man to work for," Eriksson said.

"He has his moments, it's true," Werner agreed, "but he's done so much for the company and for me personally that I have learned to ignore some of his disagreeable points and just do my work as best I can. He's originally from Bavaria and they can be quite charming people when they want to be, so it's not all bad."

"You say he's done a lot for you personally?"

"It's a long story, Lars, but he's really more like an uncle than my boss. Tonight you'll see how pleasant he can be. He loves to play the host and he's planned a sumptuous feast for us in the old part of the city."

"Hmph, it better be good," Eriksson said, "after making us watch those damned machines stuff pills into bottles for God knows how long."

Werner sighed. "I apologize again, but what could I do? The boss's orders are the boss's orders. Believe me, of all my problems with Franz, being a tour guide is the least of them." He laughed and Eriksson let the subject rest.

25

Werner turned to Goodman. "Paul, you look tired. Ready for a rest? I can arrange transportation back to the Fürstenwald for the both of you."

"No thanks. I'm fine, just a little bushed."

"Jet lag?" Eriksson asked sympathetically.

"Probably."

"Well, I'll go back with you now, if you'd like. I could freshen up a bit myself."

Goodman's head began to ache again. Why not go back and take a nap? "Yes, I think I'll accept the invitation, Lars. Peter, if you could send for a car, I'd appreciate it."

Werner hurried off to make the arrangements.

"I had the strangest beer last night," Goodman explained. "Left me with an awful headache."

"You're probably not used to these strong German brews."

"To top it off, I woke up in the middle of the night convinced someone was in my room.

"Anything taken?"

"No, my money and my watch were still there. There was nothing else of value, so I thought I must have been dreaming, except my drug dossier looked like someone had been going through it. The pages were folded back in a way I never use."

"Maybe someone was photographing it," Eriksson said jokingly.

Goodman remembered the strange clicking sound. Was that a camera? He instantly dismissed the thought as ridiculous, yet wondered . . .

"But seriously," Eriksson went on, "double bolt your door next time. Despite what our hosts say, the Fürstenwald's just a trifle too seedy for me." Looking somewhat embarrassed, he changed the subject. "I've had rather a strange request from a colleague in Berlin. I'm almost ashamed to tell you about it."

26

"Try me."

"The fellow's name is Ernst Grundig. He'd like to meet us tonight for a drink before dinner."

"Where does he want us to meet him?"

"At the Fürstenwald, since we're both staying there."

"For what purpose?"

"I'm not exactly sure; that's what makes it mysterious. He wouldn't tell me. But he is well regarded in Europe. Have you heard of him by any chance? He's at the Cardiovascular Research Institute in East Berlin."

"No, I don't know the name, but I've heard of the Institute. It was supposed to be the showcase of the country before reunification."

"Yes, that's my impression as well. Shall I tell him six-thirty then?"

"Fine."

• • •

Grundig's appearance was almost exactly opposite to that of Eriksson. Where the Dane was tall, thin and nearly bald, the German was short and stocky, with a full growth of bushy black hair. Heavy black-rimmed glasses gave him the appearance of a small studious bear.

The three of them sat awkwardly at a small table in the Fürstenwald's lounge, a comfortable room off the main lobby.

"I'm fascinated by this new drug of Peter Werner's," Grundig said after the necessary introductory small talk had been concluded. "Unfortunately, I was not invited to your research meeting so I must get my information second-hand."

Eriksson laughed. "You know Richter would never let Peter invite anyone who had anything to do with the eastern bloc, even in today's enlightened atmosphere. He probably still considers you a scientist of the old regime."

Grundig nodded. "It doesn't matter. We still have sympathizers everywhere in the country, even in pharmaceutical plants in Frankfurt, so we eventually find out what we need to."

"Sympathizers?" Eriksson said, shaking his head. "You have no 'sympathizers' any more. Paid informers and industrial spies would be more accurate terms. Rumor has it your institute is now financed by North Korea."

Now it was Grundig's turn to laugh. His dark, bushy features showed the first signs of animation since they had sat down. "Lars, what does it matter what one calls our friends or who pays them? In the final analysis, these sympathizers provide us with information that otherwise would not readily be available to the remnants of progressive elements in the East, considering the upsurge of nationalism in Germany."

"And your 'friends' have told you good things about Frankfurt 368?" Goodman interjected.

"Very good things, Dr. Goodman. Good enough for us to want to get our hands on the drug. We could collaborate with the clinical trials that are under way or being planned. And on a personal note, I could use the drug to treat one of our former senior party people who has not responded to conventional medications. Don't ask me for names, because I'm not allowed to say more. Let's just say that before the Wall came down, he was near the top of the government hierarchy and respected in the West as well."

"Do you think Richter would be impressed by such a request?" Eriksson asked sarcastically.

"I do. It would be quite a feather in his cap if his company's experimental drug was used to treat a VIP, even a Communist one. And my institute would be quite willing to give credit where it is due and if the man is not helped, no blame would be attached to the drug. We know it is still experimental. But I need your help, gentleman. Convince Richter to let me use it, or even better, give me some of your own supply. My institute will pay you handsomely for it."

"Either way Ernst, that would be a very difficult thing for us do do," Eriksson said. "Richter would probably throw us out of his office and anyway, right now neither one of us has the drug. We're more or less neutral referees to this point."

Grundig was nonplussed. "A small sample of the drug, 10 vials for example, would be worth 50,000 U.S. dollars to my institute. Think about it."

Goodman and Eriksson looked at each other incredulously. "What you are asking us to do is unethical," Eriksson said flatly.

"I'm on a difficult mission," Grundig said defensively, "so please bear with me. If you can't supply the drug, perhaps my benefactors will find someone else in the company who can. They have many contacts."

"You don't seem especially happy in your role as courier," Goodman mused.

"Yes," Grundig admitted, "that is true." He took off his thick glasses, wiped them with a napkin, and then carefully replaced them. His mood became more somber. "Shall we have our drinks now? It's about seven and I don't want you to be late for Richter's dinner. Oh yes, I know all about that. I understand he can be quite angry when people are late."

The three of them sipped their drinks quietly. Later in the car on the way to the restaurant, Eriksson was still upset about Grundig's offer. "It was as if he wanted to provoke us. He knew we would never agree to exchange the drug for money. He was as uncomfortable making that offer as we were in listening to it. I wonder if there even is a VIP in Berlin who needs it."

"Then what was the purpose of the meeting?"

"Damned if I know," said an angry Eriksson.

Israeli Security Zone, Southern Lebanon, January 4

Hands in the pockets of his winter battle jacket, Capt. Dan Tamir of the Israeli Defense Forces stared wistfully at the blue Mediterranean stretching before him, then closed his eyes and imagined himself out there on the water, sailing. Despite his fantasizing, he knew the realities of life were such that when he opened his eyes, he would still be in Lebanon—and he would be there another 50 days before he could enjoy the luxury of sailing his boat once more—or any other similar luxury for that matter.

Instead of relaxation, he had the alternate evils of boredom and combat to contend with: boredom from hours of doing nothing, interspersed with periods of sudden combat when the Shiites or their Palestinian allies decided to launch another mission against the Israelis and the Christian militia that they maintained in South Lebanon. Lebanon he could do without, Tamir decided. But Lebanon was where he was, for better or worse, in a cluster of camouflaged tents behind double rows of barbed wire in a very pretty, if chilly, setting in a very dangerous part of the world.

"Danny, are you daydreaming again?"

Tamir didn't have to turn around. He knew who it was. "Why not daydream, Ari? It's better than staring at the walls of the tent." Tamir sighed, and raked his hands through his thick black hair.

"If it's too quiet here, we can always get the terrorists to create some more excitement for you, like they did last week." Last Thursday's Hezbollah raid had caused no casualties but had been a noisy affair, nonetheless.

"Don't even joke about it. Reservists prefer boredom to getting shot at." Tamir turned away from the sea to face his intelligence officer, Lt. Ari Cohen. Ari's red hair and infectious grin made him look like a new Bar Mitzvah boy, but Tamir knew that the redhead had seen more than his share of combat in '82. Cohen was also from Tel Aviv and was also doing his 60-day reserve stint; it was only natural that a mutual bond had developed between the two of them. Except for the color of their hair, they could have passed as brothers.

"Any trouble in our part of the zone?"

"No," Cohen said. "It's quiet today. Nothing doing anywhere in the security zone."

"No booby-trapped cars? No religious fanatics who can't wait to die and go to heaven? Wonderful, but any time they want to transfer me to Nablus, I'll go."

"You better hurry," Cohen said. "We might not have troops there much longer, not after the famous handshake. But, as far as I'm concerned, the sooner we get out of there, the better."

"You're a peacenik? After all you experienced in Beirut?"

"That's just the point. What I went through can make a peacenik out of anyone."

"Or the opposite."

"True," Cohen reflected, "or the opposite!" He went over to Tamir and put his arm around his colleague's shoulder. "At least you can see the sea from here."

Tamir nodded. "The only saving grace."

"That's right, always look for the best no matter how bad the situation. Which reminds me, the C.O. wants to see you."

"When did he tell you that?"

32

"He grabbed me when I was in the headquarters tent a few minutes ago. Don't know what he wants. Maybe he found out you're not happy up here."

Tamir laughed. "Why should that surprise him?"

Cohen shrugged. "It wouldn't surprise him, but it might surprise his visitors."

"Top brass?"

"No, just the opposite, civilians. They have that Shin Bet look."

"Security agents visiting us? Wonder what they want?"

This time it was Cohen's turn to laugh. "Maybe they're here to recruit you." His prediction turned out to be accurate, as Tamir learned when he visited the headquarters tent. Standing next to his commanding officer were two casually dressed civilians. One was around Tamir's age and blond; but it was the one with a well-lined face and grey hair who did all the talking after the colonel introduced them. They were not from the Shin Bet, Israel's internal security agency, but rather the Mossad, its international counterpart.

The older agent waved Tamir to a chair. "Captain, your parents emigrated from Berlin in the '30s, correct?"

"Correct."

"They taught you German after you were born and you've kept up with it in your studies."

"Correct."

"So, you're fluent in German. Plus, you've learned English and Arabic."

"Correct again." Tamir wondered why they were asking these questions. They wouldn't have bothered making this trip if they hadn't checked out his personnel files first; they knew the answers.

"You know all of this," Tamir said impatiently. "Let's get to the point."

The Mossad agent lit a cigarette and inhaled long and slowly. "We need someone with your military background, someone who can handle himself in a difficult situation. You also have several other features that make you unique."

"I don't know whether to be flattered or not." Tamir grimaced, but his curiosity was definitely aroused.

The agent turned to the colonel. "I'm about to really flatter him. Do you think he deserves it?"

The colonel nodded. "He's a good officer," he said. "One of my best."

"Tamir," the agent continued, "not only are you a paratrooper who speaks several languages, but you are on the faculty of Tel Aviv University. You have a Ph.D. in neuropharmacology and you're an acknowledged expert in your field. Because that personnel profile is exactly what we need for a special mission, the Commanding General of the Northern Command has detached you from your unit and assigned you to the Mossad as a special liaison officer for the remainder of your reserve activity. Here are the written orders which state this."

"What do I have to do?" Tamir asked after glancing at the papers.

"You're going to have to do some traveling for us. We're not sure of all the details yet, but I think it's safe to say Germany will be one of the key places."

"When do I leave?"

"You'll be leaving your unit here the day after tomorrow. My colleague and I have some other stops to make before then, but we shall be in Tel Aviv by then to brief you further."

34

"I'm leaving my unit for good in two days and that's all you're going to tell me about what I'll be doing from that point on? I think I deserve a little more information than that."

The agent turned his tired eyes on the colonel and said apologetically through smoke rings, "I think I'll excuse myself for a few minutes, colonel. Perhaps Capt. Tamir and I can finish our little chat outside."

The colonel seemed agreeable to the suggestion. "As far as I'm concerned, the less I know the better."

Outside the tent, the Mossad agent steered Tamir toward the bluffs overlooking the Mediterranean. "I'm a lover of the sea like you are," he said "but aside from that, I suppose we have little in common. Except, of course, for our devotion to our country." He paused for a moment to watch gulls circling overhead.

"There's a lot that I have to tell you," he said, "but it's not going to come out all at once—much of it is still unconfirmed and speculative. First, since you're going to be working with me, call me Shlomo. My young colleague back in the tent is Zvi. Like you, he speaks excellent German. Now, to business. The government is very concerned at present about the chemical warfare activities of the Iraqis and the Libyans. The Americans alerted the Germans to the Libyans' plant at Rabta and the supplies stopped when the president of the German company was arrested. Now both the Libyans and the Iraqis are being more devious. American satellites have not been able to confirm the presence of new chemical plants in Libya, but their sources—and our own—indicate that something very big is being planned, perhaps even a collaborative effort between Libya and Iraq to poison large numbers of our people and disrupt any peace treaty with the PLO. Even the

Russians seem interested in preventing such a catastrophe. This is one of the reasons the Prime Minister's Science Advisory Council thought somebody with your academic background and military background might be helpful." He had smoked his cigarette down to the butt, which he now ground into the sand with the heel of his shoe.

"What the new German connection is to all this is uncertain," he conceded, "but we have reason to believe another German company is supplying the Libyans with technical help and perhaps even more. We are under time constraints from the Prime Minister's office to learn all we can and as quickly as possible. Unfortunately, our main clue to this point is quite enigmatic. A Libyan army officer who is apparently connected with this effort 'happened' to fall into our hands while on holiday in Cyprus. In his pockets were all sorts of scribbled notes. When they were translated, one phase kept being repeated. It seemed to disturb him greatly that this piece of information was now in our hands. This is the phrase." He handed Tamir a folded piece of paper.

"Frankfurt three six eight," Tamir read aloud. "What does it mean?"

Shlomo stared out at the sea for a few minutes before he answered. "I *know* it's important, but whether 368 is part of a phone number, a street address, an airplane flight number, or whatever, is unclear. That's what I have to find out. Correction. That's what *we* have to find out. Welcome to the Mossad."

Frankfurt, January 4

Franz Richter dressed quickly. He had only a half hour before the start of the dinner that he was hosting for the visiting cardiologists. He wanted to appear informal, but his habit of always wearing dark blue suits and white shirts did not leave him much room for creativity. After careful consideration, he decided informal meant he would wear a red and blue striped tie instead of his usual dark gray. Pleased with his reflection in the dressing alcove's mirror, he brushed back several stray hairs from his temples and allowed himself the luxury of a small smile. Everything considered, the meeting was proceeding nicely, and best of all, it would soon be over. No longer did he question his decision to allow Peter Werner to have it in the first place—what's done is done, he decided. Now it was strictly a question of damage control—no, that was much too pessimistic, he corrected himself—more a matter of damage prevention. Peter's enthusiasm would not be allowed to infect Goodman and Eriksson as it had Hawthorne. The Englishman was definitely a nuisance, but no more than that. Wasting time worrying about the small amounts of the drug that Hawthorne possessed would serve no useful purpose. The American and the Dane might present bigger problems. Satisfied with his appearance—and how well he had managed the meeting so far, he walked down the stairs to his car, pausing only long enough to check his private fax machine for any messages that might have come in while he was dressing. He had trained his new "friends" to contact him after business hours and, unfortunately, they had become quite adept at that. But right now the fax tray was empty. Just as well, he sighed, I don't need any more problems for awhile.

• • •

The Frankfurt Pharmaceutica dinner was held in the private dining room of a small restaurant, the Red Eagle, near the financial center of the city. Designed to give the restaurant a pseudomedieval atmosphere, the outer brick walls were covered with coats of arms dating back several centuries. The inner walls not only had similar coats of arms but suits of armor placed in corners of rooms and stairwells. Each of the several rooms—including the private dining room, had fireplaces. A variety of drinking steins sprouted from their mantels. These attempts at achieving an historical motif had certain limits, however. The waiters scurrying back and forth with serving trays were dressed in dinner jackets, rather than livery, and the table settings were contemporary.

Franz Richter was a gracious host, making a point of effusively greeting Goodman and Eriksson when they arrived in the private dining room, up one flight of stairs from the street entrance. Werner stood by his side grimacing slightly as the older man practiced his special style of backslapping.

"Peter tells me you had a most enjoyable afternoon at the plant. We're very pleased with the place."

"Very informative afternoon," Eriksson said with a straight face and Goodman echoed his remark. His head felt fine after the brief nap and he was in good spirits.

"Our plant is a model of efficiency," Richter went on proudly. "A tribute to hard work. After all, what really was the postwar economic miracle in this country? Hard work in hundreds of plants like ours. Taken together, they have resulted in a new, stronger Germany."

"What do you mean by strong?" Eriksson queried.

"Financially sound, economically secure. Surely you didn't think I meant militarily? Militarily, we are nothing. Even united we still hide behind the Americans. But that's all to the good. At least our neighbors no longer tremble in fear when our troops do a little marching here and there. Leastways I don't think they do, except maybe the Poles," he laughed loudly. "No, we are good Europeans now, isn't that so, Peter?"

Werner nodded and Richter laughed again. "And now for something to drink. Gentlemen, name your poison." He saw to it that the cocktail order was immediately attended to by a specially assigned waiter.

Drinks in hand, Eriksson and Goodman peered at the quiet street below through one of a series of exquisitely panelled windows, then strolled over to the fireplace to examine the ancient steins on its mantle.

"Richter's a bit pompous," said the Dane, "but at least he's picked an interesting little place. I hope the food's up to it. Hello there, look at Herr Doctor Stern. What an ass. He's loaded to the gills already."

Stern was last to make his entrance, his face even more flushed than usual, knees a little wobbly. By his side was a young woman, whose long black tresses covered half of her face and whose tight dress accentuated each of her curves.

Richter cluck-clucked in mock disapproval. "When I dropped you off at your hotel you were stone sober. What happened, Herr Doctor?"

"I met this old friend in the lobby," Stern gestured toward his companion, "and we had a few drinks at the bar."

"Just a few," Richter said, eyebrows raised in humorous disbelief. Stern grew more flushed while the others laughed. "Well, we won't keep our gay bachelor too long tonight," Richter said, "just in case you two are planning a nightcap." He threw his arm around the younger man and motioned for the cocktail waiter. "That is, of course, if you're thinking of joining us, which I hope is the case." The way he emphasized "hope" was clear even to the tipsy Stern.

"Certainly, Franz, that's why I'm here." He turned to the young woman, slipped her some money, whispered in her ear, and breathed easier when she smiled and departed. He then turned back to the group. "Gentlemen," he said with a big grin, "how good to see you again."

From one side of the small room to the other, Stern bobbed like a wayward top, glass in hand and face wreathed in perpetual laughter. Richter surveyed the scene good humoredly. After allowing another ten minutes for guests to enjoy their drinks, he beckoned to the waiter that they would be having dinner now. Seating his guests on either side of the table, he took his place at the head.

The appetizer was smoked trout and was accompanied by the first of several wines. "An excellent Riesling," Richter declared and toasted the success of 368. The meal began. Whenever there was a lull and the conversation dwindled, Richter would stimulate it with his own comments.

"What do you think of our little city, Professor Eriksson? How does it compare to Copenhagen?"

"Little? I wouldn't say it was little, but I like it. I do. A bit more industrial than Copenhagen. More bustling. But this restaurant is charming. This whole area is rebuilt, isn't it?"

"Yes, we had to rebuild a lot, but I think in general an attempt was made to preserve some continuity with the past. But enough talk! My soup's getting cold with all this chatter."

The meal resumed, conversation returning to comments on the weather and other trivialities.

They were well into the main course—a crisp duckling—when shouts were heard in the street below. At first, the diners ignored them, but they grew louder. Even Richter's attempts at conversation—he switched to a discussion of the comparative merits of the German versus English soccer teams in World Cup competition—were unsuccessful in drowning out the commotion. Soon the piercing blare of police horns—*ooh*-ah *ooh*-ah—grew closer. In the street someone screamed. It sounded like a woman.

Eriksson said, "What in the world's going on out there?"

Richter put down his glass, carefully folded his napkin, rose from the table with a frown on his face, and looked out one of the windows. Not satisfied with what he could see, he cranked it open and stuck his head into the cold night air. The others joined him.

"What's going on?" Eriksson asked.

"Here, come see for yourself," Richter said, scowling. The other windows were quickly opened and they all peered down in turns. The scene below had elements of a drama. Two police cars with flashing blue lights were parked at either end of the narrow street, effectively stopping the flow of vehicular traffic. Two rows of policemen converged on the center of the street from each side. Their objective was a noisy group of dissidents who had gathered in front of a small restaurant nearby and were shouting obscenities. Only ten to twenty in number, they made up for their lack of manpower by the vociferousness of their screams. Their obscenities were answered in kind from the open door of the restaurant they had besieged, which in actuality was more of a tavern. Fists were raised by the young men and threatening gestures made, but as the policemen converged on their flanks, they grew quiet. As if by signal, the answering cries from the restaurant grew more defiant, and the young lady who screamed before let loose with an encore. Bystanders watched curiously.

"Can any of you make out what they're saying?" Eriksson asked. "It looks like something political."

As if for confirmation, the group inside the besieged restaurant began shouting anew, further infuriating the gang outside. Despite the menacing looks of the policemen, they reiterated their shouts. Their voices grew louder, even as the line of policemen herded them forcefully toward one end of the street. The incident appeared over, but then one of the young men suddenly pulled out a banner from beneath his coat and began waving it furiously in the frigid air while the others applauded. A black swastika in a white circle on a sea of red danced back and forth.

"Good Lord," Eriksson gasped, "I never thought I'd see one of those things around here again." The youths began singing the Horst Wessel song, the old Nazi anthem. The police had reached the end of their patience. They swooped down on the young man with the illegal banner, throwing him roughly into one of the police cars. Brusquely pushing the others away, they broke them into little groups, raising nightsticks to hasten along laggards. With the dissidents gone, the police also withdrew. Within a few minutes, the narrow street was quiet and vehicular traffic was flowing again. Upstairs, the group at the window returned to their dinner.

"What was that about, I wonder," Eriksson mused. "Couldn't be *just* politics. Incidentally, Dr. Richter, one of those Nazi types looked like one of your workers. A good-looking blond fellow with a thick moustache."

Richter scowled. "I doubt very much that any of my people are down there. Those are our worst elements, gentlemen. In the little pub was a local group of anarchists. They're the latest fad among the young." He laughed in derision. "The Communists are too conservative for them. And in the street was an equally vociferous group, our so-called neo-Nazis. None of my plant workers could be associated with them. They're quite obnoxious."

Werner now joined in the conversation. "The neo-Nazis are only a handful," he added apologetically. "Every country has some, even France, even the United States. Isn't that right, Paul?" Goodman nodded. "They just draw more attention here," Werner concluded. "Maladjusted skinheads, that's all they are."

"Exactly," Richter said. "They are portrayed out of all proportion to their numbers. All they do is make a lot of noise and set off some bombs and terrorize some Turks and get people stirred up with memories of the past. And, of course, the leftists goad them into doing it so they can point and say, see, the Nazis are back, the Nazis are back. But the past is the past," he said emphatically. "Let sleeping dogs lie. These young idiots don't understand that. Jail the whole lot, I say." Flustered from his harangue, he threw his napkin on the table. "Come, let's return to our food and talk of happier subjects."

For the rest of the evening, the conversation remained firmly fixed on nonpolitical subjects, leaving Goodman the time to reflect on what he had just seen and heard. After mulling it over for a few minutes, he realized that the incident hadn't upset him very much at all, a reaction that both relieved and surprised him.

When they had finished eating, Richter arranged for a taxi to return Eriksson and Goodman to their hotel. Poking Stern in the ribs, he warned,

43

"Remember your age tonight, Herr Doctor. Don't get a heart attack." Unperturbed, Stern checked his watch and started to leave. Richter laughed. "One moment, Dr. Stern. Before you go, I need to have a word with you. I will see the rest of you in the morning, gentlemen, when it's back to business. Frankfurt Pharmaceutica hopes you've enjoyed the dinner."

Eriksson and Goodman waited at the street entrance for their taxi, the Dane still engrossed with the disturbance he had witnessed. "That's what 1932 must have been like, right-wing and left-wing fighting in the streets. Unbelievable to see it now played out again in front of our eyes, don't you agree?"

"It was an isolated incident," Goodman replied, "nothing more."

"That's what Richter says." Eriksson's voice was reproving. "And I still think I saw one of those street fellows in the factory. Don't you remember him from our tour? I think he was one of the chaps coming out of the guarded room."

"Yes, I remember seeing the blond fellow at the plant, but I didn't see him in the street tonight. In this instance at least, I think Richter's right."

Eriksson snorted. "Perhaps, but I remember a conversation I had with a cardiologist in Munich a few years ago. A Dane married to a German and living in Munich for over ten years. We were at a post-symposium party in one of the big Munich beer halls. Oompah band, lederhosen, little hats with feathers—the whole works. Well, after awhile the place grew quite lively, what with all the beer flowing and the noises getting louder and louder. I turned to my friend and said, 'You know, if I didn't know better, I'd think they won the war.' I expected him to grow very defensive about the Germans—after all, he was married to one and lived there quite contently—but instead he

44

became very serious and looked around at the crowded room. 'For the right leader, they'd march again, all of them, and don't you ever forget it.' He meant it, too."

Goodman said nothing and in a few minutes the taxi arrived.

Frankfurt, January 5

Only Richter, Werner, Goodman, Eriksson, and Stern participated in the conference the morning after the dinner at the Red Eagle. Instead of the larger room with its panoramic view, they sat in a smaller, windowless room adjacent to Richter's office. The general director of the company was in an ebullient mood and as they settled in, said, "I see everyone, including Dr. Stern, has recovered from the food and drink consumed last evening."

Stern's eyes widened. "Extremely nice dinner," he said mischievously, "but I don't remember drinks being served. Did I miss something?"

"You ate and you drank and God knows what else you did. God and that young lady, though she didn't seem like a lady." Richter wagged a finger at Stern and roared with laughter. Then turning to the day's agenda, he quickly sobered and began speaking. "Today's business is 368 and after yesterday's presentations, I'm sure there is a natural inclination to proceed quickly to testing the drug in as many patients as possible. I must admit to you that there is some difference of opinion within the company on how to proceed in this matter. For my part, I believe we must proceed slowly so as to minimize adverse publicity and even damage suits if 368 proves ineffective or—God forbid—dangerous."

"I disagree," Werner said forcefully, "I think more clinical trials should begin in our country at once. Hopefully, they can also be started soon in the United States, Britain, and Denmark, and the sooner the better."

"So there you see the dilemma, gentlemen. My research chief and I are at odds, friendly odds, of course. Let me get some idea what you think. Walter?"

"It's true, as Peter has implied, that there have been no dangerous side effects in the animals or normal volunteers that we've tested, but our group of patients with heart disease is very small and we have yet to be certain about toxic side effects in them, despite the generally good results. I would be in favor of continuing the studies I've been conducting in Hannover and holding off on foreign trials."

"What do you think, Professor Eriksson?"

"I agree somewhat with your point of view, Dr. Richter, and that of Dr. Stern. The place to do the studies is in Germany, or possibly in Denmark, especially given the difficulty of getting quick governmental approval for new drug studies in Britain or the United States."

"That's true," Richter commented, "Denmark would have less problems than the other countries. I know Hawthorne has had a devil of a time getting approval in London and he was as enthusiastic a supporter of expanding clinical trials when he visited us as Peter is."

"But he will have approval soon," Werner interjected. "I heard from him just the other day, and things are moving along much better now."

"Is that so?" Richter said in obvious surprise. "Why didn't I know about it?"

"I'm sorry, Franz, but I didn't have a chance to tell you, what with all the preparations for this conference."

Richter's face grew redder as he stared at his research chief. "You must keep me informed of these things!" he burst out. "What good is having a general director if no one tells him what's going on?"

"Sorry," Werner repeated glumly.

Richter shook his head in disgust, then made an effort to calm down. Turning to Goodman, he said, in a more subdued voice, "What do you think, Dr. Goodman, can you receive prompt approval or is Professor Eriksson correct? Will you have difficulty?"

"We will have difficulty. Our Food and Drug Administration is still slow in these matters," Goodman said. "But since we have no plans to give this drug to pregnant women or very young people and since your animal studies and normal human volunteer studies are all in order, I see no reason why they would hold us up too long. Especially if we are going to use 368 only in very sick patients."

"The cases so far have been confined to the seriously ill," Werner said.

"But should it be confined to those kinds of cases? Is it fair to the drug?" Goodman wondered.

Werner shrugged. "I have asked myself the very same question, Paul, but at the moment that is all company policy will allow us to work with. I have a feeling that Dr. Stern's patients were too sick to show better results. Isn't that possible, Walter?"

"It is possible," Stern conceded.

"I'm concerned that if we only take patients like these," Werner continued, "and word gets out that we have failed, 368 will be discredited."

"That's my responsibility," Richter replied.

Werner persisted. "But why not give permission to work with patients who are not necessarily refractory to existing drugs?"

"No. Not yet," Richter said. We take only the sickest; that way we *can't* get in too much trouble if it fails. I say we *go slowly*, gentlemen. If you are willing, we will fund you; don't concern yourself about that even if the studies don't start for a year or two. Not extravagantly, of course, but enough for clerical and technical help, etc. But I say we still only use it in the kind of patient Dr. Stern has already used to begin with."

Aside from Werner, no one voiced any further dissent on that issue.

Richter was pleased. "Good. And until we have some better human confirmation of our animal studies, let's not talk too much about this. We don't want our competitors to know more than they have to."

"Is that much of a problem here?" Goodman asked.

"Yes, it is. Not only from West German concerns—who are very aggressive—but also from some of our new 'friends' in the reunified areas in the East. There are still some hard-core Communists there who would just love to claim this sort of discovery as one of their own. I have trouble trusting many of their so-called 'scientists.' "

"And what about your initial animal work?" Eriksson said, quickly switching the subject after a quick look at Goodman. "Will you publish that?"

"Of course," Werner insisted. "I'm preparing the manuscript now. As long as the structure of the compound is known only to us, there can be no easy duplication."

"Well," Richter said somewhat hesitantly, "we may have to be careful how we present the material even there."

"When the manuscript is ready, it will be submitted," Werner said defiantly.

Richter smiled and held up his hands in mock surrender. "The lag time between submission and publication should give us a big lead anyway. Just let me see it before you send it in. No more surprises, please."

Werner nodded.

"Then as I understand it," Eriksson said, "you have decided to proceed slowly. Is that correct?"

"Correct," Richter said.

"In the meantime, I think I can get government approval for clinical studies in Denmark. You have no objections?"

"Excellent idea," Werner said, looking at Richter, who did not object. "Furthermore, we should use this meeting to discuss what future trials will consist of since, hopefully, we will eventually be starting them in all your countries."

The meeting continued with the discussion and planning of detailed protocols for the administration of 368 to additional patients. The final item discussed was the financial arrangements for supporting these future research studies.

"You presume that the results of the trials will be favorable to 368," Goodman said to Richter when the funding arrangements were discussed. "What if they aren't? Do we still have a free hand in the performance and reporting of these studies, or must we adhere to the wishes of your company?"

"What if I told you, Dr. Goodman, that no results were to be published without my approval?"

"Then I would not participate in the study," Goodman replied without hesitation.

"Nor would I," added Eriksson.

Richter laughed. "Then I certainly would never say such a thing. No, you may have no fear, gentlemen, you are in no way beholden to us for anything more than a fair trial of the drug's worth. Whatever scientific reports result from these trials will be entirely free of censorship by any employee of Frankfurt Pharmaceutica, of that I give you my word. I hope that will be enough!"

He left it up to Werner to summarize the day's discussion, then announced there would be a final meeting on Wednesday morning to make arrangements for future meetings and to formally conclude these initial exploratory sessions.

When they were in the hall and the others had walked on ahead, Werner took Goodman aside. "Are things going all right so far?"

"Fine. I'm just sorry I couldn't give you more support this morning, but I'm naturally skeptical when it comes to new drug claims."

"I understand, don't think about it as something personal. Anyway, I may have some information soon that might make you change your mind. I'll write to you in New York when I have a little more data."

"Nothing you want to talk about yet?"

"Exactly, too premature."

"I have a sneaking suspicion you're *not* talking about animal studies. You're doing your own patient studies, aren't you?"

Werner put his finger to his lips. "Don't even talk about things like that. You want my head on a platter?" He winked broadly, and gave a twist of his head so the platinum blond hair fell back across his brow.

"I thought Stern was handling all that for you."

"Stern was Richter's choice, not mine. I have other friends in the clinical area. But I'm talking too much already. Now what do you say to a quiet evening at the Werner home? Sample my wife's cooking and see my wonderful kids? We don't live very far from here."

"I'd be delighted."

"Good. Let me get my briefcase and and we'll leave immediately." Werner left Goodman in the hallway while he quickly returned to his office. Stuffing papers into the already bulging briefcase, he wondered what he said or had done to make Richter so upset.

Today had been even worse than usual. Frowning in annoyance, he hurried back to his guest.

• • •

Werner was a very cautious driver, keeping in either the right or middle lanes of the three-lane highway. Glancing repeatedly at the speedometer, he made sure his speed never exceeded 70 kilometers/hour. Aware that Goodman was staring at him quizzically, he smiled that warm smile of his again.

"Too many idiots on our roads, Paul, and either no posted speed limits or no one pays attention. This way I know I'll get home in one piece, or at least I'm doing whatever I can to see to it. I'm afraid I'm the complete stereotype of the family man. At times, I'm afraid I must be terribly boring. It's a sign of self-control that I haven't bombarded you with family pictures already."

"I've driven with slow drivers before; it's nothing strange. Only I've never been passed by a Volkswagon with four nuns in it before. That's really slow driving."

"No! I don't believe it. Where?" Werner looked to the side and in front to see the car full of nuns, then realized that Goodman was only joking. He laughed. "I deserve it. But when I was younger," he said more soberly, "I crashed up the family car pretty badly and nearly killed myself. I'm not about to do it again." And with that, he settled back at his comfortable cruising speed.

Traffic thinned out as they left the main highway and branched off onto another highway that led to one of the small western suburbs. From there, it was only a short drive before they turned into a pleasant street lined with a row of substantial brick houses, all having red tile roofs, but not much in the way of lawns.

Werner slowed down as they approached one of the homes halfway down the street. When he beeped the horn, a boy and a girl, so bundled up that only their fiery red cheeks were visible, ran out of the backyard toward them.

Werner parked the car at the end of the driveway and embraced his two children for several moments. The children had carefully saved some of the morning's snow and now proceeded to gleefully bombard their father with a stream of small snow pellets. Werner cringed in mock fear and cried out, "Help. They got me. Oh, they got me this time. Now I'm done for." And with a few loud gasps and much eye-rolling, he feigned collapse on the hood of the car.

"Oh, Papa, stop joking," said the girl.

"Yes, stop right now," the boy insisted, and with his boot gingerly prodded his father in the leg.

Goodman watched with amusement as Werner suddenly sprung up and grabbed each of them in turn. A bear hug and a kiss to the cheek was their reward for good snowball marksmanship.

"Where's your mother?" he said, releasing them.

"Inside, cooking," they answered in unison, and ran inside.

"Good kids," Werner said, beckoning Goodman to follow him. "Now once we go in the house, no more talk of business—that's my wife's rule—and it helps keep a sense of balance. Agreed?"

"Fine with me."

The interior of the house was small by American standards but looked comfortable. After taking Goodman's coat, Werner invited him to relax on a soft sofa in the living room and went to get his wife. Goodman sank into the soft pillows and put his head back. From the corner of his eye, he saw the two children eyeing him from the entry alcove. Without their winter garments, they were less roly-poly. Unlike their father, they were both brunettes with dark eyes. As soon as Werner returned with his wife, it was apparent the children had inherited her coloring. Werner's wife was tall and handsome with dark brown hair that contrasted sharply to the platinum blond of her husband's. Goodman rose to greet her.

"Welcome to our home," she said after her husband had introduced her. "I hope you like roasts." She wiped her hands self-consciously on her apron.

"Thank you for inviting me. Your children are adorable."

"Oh, they're all right," she replied, slightly flustered.

"Please sit, Paul, I'll get you a drink. Trudi, would you like to join us?"

His wife shook her head. "In a few minutes, perhaps. Just a few last-minute touches and I'll be ready."

"What about Johanna?"

"She's still upstairs with her books. I'll tell her to come down. I just hope . . . "

"That's my sister-in-law," Werner interpreted. He explained to Goodman, "She's a wonderful girl, but very opinionated. I don't think she cares very much for your country's political leaders. Trudi, tell her to come down and have a drink with us and put those damn books away for awhile. And, children, I think your father can use one more kiss before you go and play a bit before dinnertime." The children ran to him while he sat on the floor. He positioned one on each thigh. "See, Paul, this is why I'm such a slow-poke on the autobahn. My kids and wife are more important than idiots who want to race with you." He kissed them forcefully on

54

their foreheads. The smaller one threw his arms around his father and whispered into his ears so that the older man laughed. "Yes, you are right. A guest for dinner means special dessert—but now go and play for awhile. Skoot." They both ran off squealing happily. Werner watched them go, then turned to his guest. "Well, what will it be, Paul? Scotch, bourbon, wine?"

"Scotch on the rocks."

"What else? The true American drink." He prepared the drinks at a small bar at one side of the room. While his back was turned, a very attractive young woman entered the room. Of medium height, she had shoulder-length blond hair and a full but not heavy figure. Wearing a simple yellow turtleneck sweater and plaid skirt, she looked at Goodman with curiosity.

"Hello," she said softly.

"Hello," Goodman replied. Her hair was not as light as Werner's, and the darker hue complemented her blue-gray eyes.

Werner turned at the sound. "Ah, Johanna, you've torn yourself away from your studies. Wonderful. Come and join us. Paul, this is my sister-in-law, Johanna Bauer, the professional student." He gave Goodman his drink and joined him on the sofa after pouring his sister-in-law a glass of white wine.

Johanna sat opposite them in a chair. When she crossed her legs, Goodman felt his pulse quicken at the brief flash of pink thigh. He flushed slightly.

Werner's wife returned from the kitchen minus her apron. She placed some cheese and cold hors d'oeuvres on a coffee table near the sofa while Werner poured her a glass of white wine.

"Well, Dr. Goodman, what do you think of Peter's wonder drug?"

"Trudi!" Werner said, surprised. "I told Paul that we don't talk business at home and here you're breaking your own rule."

"Tonight, I'm making an exception. For the last year, I've heard about nothing but the wonder drug—the famous 368. I want to hear what an impartial observer thinks of it."

"Even I've heard about the wonder drug," Johanna said, "and I have no scientific interest at all to judge its merits."

"My sister is studying for her doctorate in modern European history at the University of Heidelberg," Trudi explained, "and so she manages to avoid scientists. Though it doesn't seem to hurt her social life."

"If you hadn't married me at such a young age, you could have the same life," Werner said jocularly.

Trudi smiled, then turned once more to Goodman, repeating her question.

"To be frank with you all," Goodman said, "I take a very cautious approach to any new drug. My attitude is one of skepticism; in other words, I have to be convinced. As Peter well knows, claims for drugs and other forms of therapies are easy to make, but hard to prove."

Werner nodded energetically. "I don't disagree with that approach at all. But I just feel . . . in my bones, if you will . . . that this is a fantastic opportunity."

"Dr. Richter is less enthused, though," Trudi joined in, "and he is the head of the company."

"Don't misinterpret Franz' attitude," her husband replied, "he's all for our work, but he doesn't want to get egg on the company's face with false claims . . . "

"Still, I wonder sometimes if he's jealous of your work."

"Nonsense. He's almost like a father to me now. He's happy to see a chance for me to achieve a breakthrough like this. But he's just . . . cautious. And besides, he must think of the entire company, not just research and development."

Trudi played with a slice of cheese before responding. Her tone was gentle, but persistent. "Perhaps I should explain, Dr. Goodman, that this company was founded by Peter's father and Franz Richter right after the war. When Peter's father died several years ago, Richter assumed the directorship."

"My father and Franz were army doctors together," Werner added. "They became very close during the war."

"Not our most glorious period of history, I'm afraid," Johanna interjected somberly. "But then again, Germany is not the only country with fascist policies. Your country, Dr. Goodman, seems to be as pig-headed as we were in the '30s."

"Oh, Johanna, don't start in on your tirade," Trudi said harshly.

"Actually, Dr. Goodman," Johanna went on ignoring her sister, "there is great popularity amongst all age groups in criticizing you for Vietnam and your Central American adventures, but our older citizens have a great reluctance to discuss the crimes of the Nazi era. For a while, any talk about the Nazis used to be avoided like the plague, even in this house."

"Johanna, Dr. Goodman has other things on his mind than your theories about politics."

"That's all right, Trudi," Goodman said, "I'd like to hear what Johanna has to say."

Trudi's eyes widened and she shot her sister a reproving glance.

Johanna didn't appear to notice. "Peter told us you were Jewish, so you must have had mixed feelings about coming to Germany."

"You're right, I did have some," Goodman admitted.

Werner cleared his throat noisily. "That was another generation, one that we have no control over and are ashamed of. By we, I mean the younger generation, the three of us in this room, our friends and colleagues. We repudiate all that our parents' generation stood for."

"You don't really repudiate your father" Johanna said. "In fact, you idolized him."

"My father was in the army," Werner said wearily, "and he was no better or worse than the rest, but we never spoke much about it. His generation was very closemouthed and my father was the same way."

"And what of *our* parents?" Trudi asked her sister. "Have you ever sat down and talked with them? They had nothing to be ashamed of."

"My father and mother are very old, Dr. Goodman," Johanna explained. "My father wasn't even in the army until the war was almost lost. But they were silent like the others, out of fear or of acquiescence, no one can say. Even now, they don't—or won't—say how they felt fifty years ago. It's the same old business—we didn't know what was happening, etc."

"What's the point in digging it up again?" Trudi said.

"Many people in this country don't want to, but I do," Johanna persisted. "Because there are remnants all around us, ex-Nazis are still in positions of power. You should see the SS reunions with their torchlight parades and forbidden party songs."

Trudi scowled. "That's a lunatic fringe, old men with their dreams of hate. And despicable young men who emulate them."

"Johanna's right about that," Werner said. "Not all of the Nazis are in South America and not all of them

58

are old. And what's worse, they seem to be enjoying a resurgence in Leipzig and other cities in the East. Some call them harmless skinheads, others aren't so sure. But enough of that." He seemed embarrassed. "Trudi, go and see if dinner is ready. We can talk later." That ended the conversation.

The roast was fine, but Goodman ate sparingly. He was distracted by Johanna. He couldn't help glancing at her across the table. She seemed to sense his interest. She turned to her brother-in-law after dinner. "If I can be excused from kitchen duty, perhaps I can show Dr. Goodman around our neighborhood?"

"Go ahead," Werner insisted. "I'll help Trudi with the dishes. But give the children a good night kiss in case they're in bed when you get back."

The streets were dark and quiet. Goodman and Johanna walked together, hands thrust in their coat pockets, enjoying the night air. Great clouds of breath hung in front of them.

At a store front, they stopped and Johanna leaned against the window, nose to the glass. "I can never resist looking at the toys. The kids must love to visit this place." Then, more seriously, "Being a student of history has its good points and its bad points; you learn more about things than you thought you wanted to know, like Nazis and concentration camps. Unfortunately, my country has had a very violent past. We still have problems, though of a different kind. That's why I'm active in the Green Party. We are for a clean environment and a neutral foreign policy. By the way, I dislike your government very much," she added defiantly, "and I must also say I find many Americans quite disagreeable personally."

Goodman grinned. "As long as that doesn't include me, I won't argue the point."

Johanna continued as if she hadn't heard him. "Sometimes I feel the radical students have a point, that western society is too materialistic, too greedy, but when I see what they propose instead, I'm not very enthusiastic. All the violence in the world, the senseless killings and kidnappings in the name of God or some other holy cause. Not for me. I stand for democracy and decency and freedom. We Germans have had little enough experience with those things, there can be no tampering with them now. I'm especially concerned about our new citizens in the East. They have no democratic traditions at all and seem ripe picking for demagogic leaders. Some of my colleagues say I worry too much about the future of our nation, and everything will be fine. But I don't care what they think of me. I'm an independent sort of person and I like it that way."

"That I can see."

They stood silently in front of the window full of toys, their breath forming mist on the glass. She turned away from him for a moment, then cocking her head to one side asked him, "Are you married?"

"No, but I was once."

"I came close once myself."

"No strings now?"

"Some." She drew a cartoon face on the misty window with a wide mouth and big ears. "How long will you be in Frankfurt?"

"A few days more."

"Would you like me to show you some of the attractions?"

"I thought you didn't like Americans."

"I'm trying to be friendly, for Peter's sake," she said defensively.

Goodman shook his head emphatically. "No thanks, I'm not a charity case. The research meeting's due to end at noon, and I'll tour on my own after that."

Johanna shrugged. "Suit yourself," she said coldly, ending their conversation.

When they returned to the Werner house, Goodman said, "I don't understand you. Why did you bother to take me on this after-dinner walk?"

"I told you. I'm trying to please Peter. Trudi's very much in love with him and she wants me to be a good sister-in-law. But it looks like I didn't do a very good job tonight."

"No, you didn't."

"I tried, but you Americans are all the same. Arrogant and domineering."

Goodman rolled his eyes upward in disbelief. "Good night, Miss Bauer."

London, January 6

The weather report on BBC-1 had predicted snow, but the independent radio station that Dr. Harold Hawthorne also listened to predicted only more rain. It was no surprise to Hawthorne when the snow failed to materialize and rain fell instead. "I don't trust that new chap at the BBC anyway," he told his wife while he dressed for work. "The old one was reliable, but not this young know-it-all."

His wife was less wrought up. "What's the difference, dear? You see for yourself it's raining. Why do you need someone to tell you what you can see for yourself?"

Hawthorne could not follow her logic. "That's not the point," he said, but then decided it wasn't worth explaining. He loved his wife, but he had long ago decided that her thought processes were simply not on a par with his. So be it, he had concluded, there are more important things to a marriage, as their 27 years together attested to.

"I'm off now," he announced with a final tucking in of his shirt.

"But you haven't had any breakfast."

"I'll get something at the hospital."

"That's the third time in the last fortnight that you've missed breakfast. If I didn't know you better, I'd think you were having an early morning affair with your secretary."

Hawthorne put on his eyeglasses and stared at his wife. "Are you daft? That woman makes you look like a beauty queen." Throwing his raincoat over his arm, he bolted for the door before she had time to react to his left-handed compliment. "Call you when I get in, as usual," he yelled back over his shoulder. His wife required daily reassurance that he had arrived safely at work. One of her minor idiosyncrasies.

Hawthorne's route to the hospital never varied, through fair weather or foul. Walk to the tube, take the tube to the nearest station to the hospital, walk to the hospital. He needed the walks, he told himself, ever since he had noticed his abdomen begin to bulge ever so slightly. Must have more exercise, he had decided, and for the last year had walked as much as he could, spurning other modes of transportation whenever possible. Today he was in a hurry: he had the final handwritten touches to put on the first draft of his new drug study and he wanted to finish the rewriting by noon so that his secretary could type it over in time to take it home and read through it once more.

He knew the drug called 368—the company still hadn't decided on a name—was the best of its kind that he had worked with in the last 5 or 10 years. The sooner he got the study into press, the better. The squabbling at the German pharmaceutical company between Richter and young Werner was none of his concern. He was a scientist; he did his work and published it. That's why they had given him samples to work with. Why Werner did not trust his boss was not a problem for him.

Halfway between the tube station and the hospital entrance Hawthorne had to traverse an undergound pedestrian passage, a "subway." He was actually relieved to enter the dark tunnel because it gave him a chance to put down the cumbersome umbrella for several minutes. No sooner had he lowered the umbrella than he felt someone tap him on the shoulder from behind. He turned quickly to his left and immediately had his arm pinned to his side by someone he couldn't see. The man who hit him he did see—for a brief moment, that is. The man was dark, he would

63

remember later, but not an African, more like a Pakistani. The man said nothing, simply smashed the metal pipe he was carrying into Hawthorne's abdomen until the older man's breath gave out and he lay gasping on the ground, the wet paving stones cold against his cheeks. Methodically, the metal pipe crashed down over and over on his body, sparing his head but not his right arm, which he used to protect his face. Almost as an afterthought, the assailant reached into Hawthorne's jacket pocket for his billfold. By this time, there was no further resistance from the professor.

The two men ran off to the far end of the subway tunnel, while Hawthorne, half unconscious, moaned in pain. The assailants had followed their instructions to the letter: avoid killing him, but put him out of action for a long time and be sure to break his right arm.

Israeli Security Zone, Southern Lebanon, January 6

After he packed his gear into a duffel bag, Tamir lingered for several moments with Ari.

"I still can't believe you're really leaving Lebanon," the redhead said, looking upset. "You've only been here 10 days and you're off already."

"What can I do? Orders are orders."

"How long will you be gone?"

"Don't know, but if I don't see you for awhile, good luck." He shook the junior officer's hand firmly, slung the duffel bag over his shoulder, and walked quickly to the helicopter pad. He didn't enjoy goodbyes.

The helicopter left the Israeli security enclave and crossed the Lebanese border into Israel near Rosh Hanikra, whose exotic grottos Tamir had visited several times. While the pilot chattered on about his newborn son, Tamir gazed down at the scenery of northern Israel, the beaches, fishing inlets, villages, vineyards, and farming cooperatives. The thought of this land and its people being exposed to clouds of deadly nerve gases seemed unbelievable, but he knew it could happen. Tamir was thirty-three; a fourth-generation Israeli. Both his father and grandfather had fought in its wars. He had lost an uncle in the storming of the Golan Heights in 1967. His two older brothers had served in the military ahead of him, one injured in a 1985 clash with infiltrators.

The helicopter flew over the port of Haifa; ships of the U.S. Sixth Fleet were everywhere. A huge aircraft carrier was the main attraction, its deck covered with warplanes. On his reserve duty two years ago, he had participated in joint landing maneuvers with marines from the Sixth Fleet; they were excellent fighters. They were also legendary drinkers.

Shore leave for thousands of sailors and marines meant Haifa would be busy tonight. The bars would be open late. Nightlife was something Tamir had not enjoyed for some time. He had broken up with his latest girlfriend—a graduate student—a month before his reserve duty began. Things had been quiet since then on the romantic front. "Relationships" were not his strong point, that he knew. His two brothers already had children and his parents were pressuring him to have his own family. He wondered if he would have time to see them before leaving the country. He'd have to get in touch with them in any event; if he missed his weekly telephone call they'd worry that he had been wounded in Lebanon. While his mind wrestled with the best way of explaining to them that he would no longer be on duty in Lebanon, he dozed off.

When he awoke, the helicopter was landing at an airbase outside Tel Aviv. A jeep and driver were waiting to drive him into the city. Tamir was surprised to find that the Mossad office that he was taken to was in a small building just off Dizengoff Street, the main shopping boulevard in the city. He had expected a more secluded spot, one that was less "businesslike." Shlomo apologized when Tamir entered his office. It was small and cramped. Zvi sat in a chair against a wall.

"Sorry for the tight quarters. No fringe benefits in this agency, but what we have to show you won't take long. Then you can go to your flat and get out of your uniform."

Tamir looked around him. On the wall behind Zvi was a high-detail map of Libya. Shlomo seemed amused as Tamir studied it.

"Recognize any names?"

"Some," Tamir answered. "I just had a sinking feeling that I could be visiting there one of these days."

Shlomo made a face. "You're looking too far ahead. I promised you Germany, not Libya. Let's hope it never comes to that."

He opened up a desk drawer and took out a manila folder. "Look through this. It's from the Libyan officer I told you about. It will give you a chance to use your skills in Arabic. Maybe something will ring a bell in that scientific brain of yours. Zvi and I won't bother you for awhile." He went back to work on his desk while Zvi read a newspaper.

Tamir leafed through the material in the folder. Most of it appeared to be personal effects: photos, letters, etc. He put it aside for the moment. He was sure the Mossad translators had little trouble with that. There were also voluminous notes on chemical processing; here is where his skills could be used.

Tamir read through the notes carefully until he found something that rang a familiar bell. "Shlomo," he said soberly, "there are two formulas here. One's for a compound whose general structure suggests it stimulates the electrical contraction of certain cells, like heart tissue. I don't know why the Libyans would be interested in it. The other formula is for a compound that would block the action of cytochrome oxidase, a necessary enzyme in all living cells. It's similar to cyanide. My guess is that it's more lethal and could be distributed as a fine aerosol spray with little difficulty, but it's only a guess and I have to confirm it."

"What's the best way to do it?"

"I'll make some discreet inquiries at the university this afternoon. That is, if you agree."

A somber Shlomo nodded. "Of course. What do you make of the other items?"

Tamir glanced through the letters. The army officer had a wife and two daughters—and a mistress in London who signed her letters "B." There was also a photo taken in front of Harrods: a darkly handsome man of about 50 with a thin moustache and a rather plain, colorless woman with hair swept back in a bun, most likely "B."

"The Libyan certainly didn't pick out a good-looking woman for his romantic fling in England," Tamir said thoughtfully. "Isn't that a bit odd?"

"Our thoughts exactly. Unfortunately, we have no return address and no name, just the letter 'B.' "

"Couldn't you learn more from the Libyan himself?"

"He died soon after he came into our hands."

"Interrogation too rough?"

"We only do what's necessary," Shlomo said carefully, but his younger companion was more candid.

"I was part of the team that captured him," Zvi said. "We were no rougher than usual, but state security sometimes requires special procedures. Remember, these bastards wouldn't think twice before dropping a cannister of poison gas right in the middle of Tel Aviv."

"It really doesn't matter to me how he died," Tamir said.

"He actually died a natural death," Shlomo continued. "He must have had a heart condition he was unaware of."

"So there's nothing more to be learned from him?"

"Nothing."

"Anything more for me to do here?" Tamir asked. "If not, I'll go to my flat to change clothes and head over to the university."

"Call me tonight at this number," Shlomo said, and handed Tamir a card. "We'll decide then on our next step."

When the army captain had left, Shlomo turned to Zvi. "Well?"

"I think he'll do fine. You've told him the minimum you had to. He accepted it with only a reasonable amount of complaints. I agree that there was no point in going into the Goodman business. Not until we're sure, isn't that what you always tell me?"

Shlomo nodded. "Not until we're sure."

Frankfurt, January 6

Goodman began his day by saying good-bye to Eriksson, who was returning to Copenhagen.

"I've very much enjoyed the opportunity to spend this time with you," the thin Dane said in the hotel lobby surrounded by his luggage, "especially since we seem to share so many similar views, at least in medical matters. I wouldn't be too surprised if we're all called together soon to actually proceed with the clinical studies that we discussed yesterday. Peter Werner's a persuasive fellow and he'll eventually get his way with Richter."

"I'm not so sure about that, Lars."

"Well, if I was a wagering man, I'd bet a few coins on it. But that's neither here nor there. Incidentally, I was right about one of those neo-Nazis being a company foreman. I spoke to Dieter Hoffman, Peter Werner's assistant. Apparently the whole plant's buzzing about the incident we witnessed. According to Hoffman, the fellow's name is Hans Meyer, and he's one of Richter's fair-haired boys. He not only works on 368, but some other projects which are very hush-hush around the plant. In any event—and here's the part you might be interested in—the word is that this morning Richter told Meyer to take a few days of holiday until this meeting was over. Dieter says that Meyer and his girlfriend went skiing. When Werner heard about it, he went into a fit. Couldn't understand why Meyer's actions were tolerated, much less protected. One more thing for them to squabble about."

"Strange business," Goodman mused.

Eriksson nodded. "Very strange. Well, I'm off. Until we meet again, here or wherever . . . " He shook Goodman's hand energetically and motioned to the doorman that he was ready for a cab.

It wasn't long before Goodman was saying a second good-bye. He had just arrived in Werner's office when Richter peeked in, a wide smile on his cherubic face. "I thought I would find you here. I'm off to Stuttgart on business and I wanted to thank you for joining our little meeting. I know we can count on your continued good advice and, hopefully, future assistance as well."

"I'll do what I can."

"What more can we ask? Peter, I shall see you tomorrow. Let's put that other matter behind us, shall we?"

"I'll try. Have a good trip, Franz."

When he had gone, Werner smiled ruefully. "A whole day without the boss around. Normally, that would put me in a good mood."

"But not today?"

"No, too much bad news. First I had a disagreement with Franz about some personnel matters, shall we say, and then I had a telephone call a few minutes before you came in that is really guaranteed to make my day miserable."

"You'd better explain."

"Hawthorne's wife called from London at his urging. It seems he's in the hospital; he was beaten and robbed early this morning. A broken arm. Can you imagine that?"

Goodman had met Hawthorne at several heart meetings, and he remembered him as an older man, very gentle, likeable, and soft-spoken. Just the type to be selected by muggers. "I thought that sort of thing only occurred in the United States; London always seemed more civilized than our cities."

"The police think his attackers could very well have been foreigners, which may explain it. Immigrants apparently are causing a lot of problems in Britain.

71

Poor old Hawthorne, he won't be back at his work for weeks or even months."

"At least he's alive, it could have been worse." Goodman remembered one of his colleagues who had been stabbed to death in a similar attack.

"Of course, I know that," Werner said glumly, "but there's more to it, Paul. I'll tell you about it on the way to the lab. Hoffman's waiting for us."

As they walked through the corridors of the building, Werner explained. "Hawthorne's been a firm believer in 368 from the beginning. When he visited us in the spring, he was as enthusiastic as I was about the drug. This was only on the basis of the animal studies, mind you, but still, he saw its potential. We gave him a small supply of the drug for use in normal volunteers and he sent the results over to us last month. They were most encouraging. Absolutely no side effects. Here's the irony—he began studies on selected patients with arrhythmias. He's used up his supply and now he needs more. And best of all, as I told Richter yesterday, he had finally gotten government approval." He stopped, looking behind him to make sure no one could hear. "And I don't mean hopeless cases either. Run-of-the-mill problems, the kind Richter has told us to avoid, and he has excellent results and and was writing them up. Now that has to be put off for God knows how long. So you see why I'm upset."

"I thought you weren't going to take any short-cuts with 368."

"This drug's too important to let a stick-in-the-mud like Franz ruin it."

Goodman remembered an earlier conversation with Werner. "Are these the studies you were going to surprise me with?"

Werner nodded. "They were. Plus some others that are even further advanced, that no one here knows about. Studies that give results far better than Stern reported. You see, it's the only way to make these dunderheads see the value of the drug . . . and convince skeptics like yourself." He walked on, hands clasped behind his back, still muttering about Hawthorne's accident.

In his laboratory, he went through an elaborate ritual of opening the safe where the vials of 368 were kept. "Richter insists on strict security," he explained. "He's afraid our competitors will steal it. He's got a mental block about the treachery of Communists— and even ex-Communists—so we humor him." Extracting a vial from the safe, he carried it into the animal experimentation room where his assistants were working on the carcass of an anesthetized dog. Hoffman's team was as efficient as ever, but nothing they would show Werner that morning—not even dramatic results in provoking and treating malignant arrhythmias in an isolated beating heart—could cheer him up. He was disconsolate. "To top it all off, I found out that one of our top workers—Hans Meyer—is one of those neo-Nazi thugs. I want to talk to him about it, but he's taken off on holiday. I did manage to do one good thing, though. I convinced Johanna to stop being so discourteous. I took it upon myself to arrange for her to meet you for lunch. Don't protest, just humor me. Maybe you'll talk Johanna into going dancing with Trudi and myself tonight. I don't want your visit to end on a sour note. That would make me a lot more cheerful."

"You really like to scheme," Goodman said, shaking his head, "but you did promise me a visit to the nightclub with the swing band. I suppose I shouldn't leave Frankfurt without seeing it."

"What's today? Wednesday, just the right day. I'll work on Trudi, you work on Johanna. See, I'm feeling better already."

Werner had arranged for Johanna to be at a small cafe on the Berlinerstrasse not far from Goodman's hotel in the old part of the city and close to the Romer, the town hall.

When Goodman arrived, she was already there waiting for him, sitting at a corner table, reading a book, her ample bosom accentuated by a clinging sweater. One leg was crossed over the other so that the skirt had risen above the knee. Just as he had the night before, he felt a rush of blood to his face when he stared at the pink of her exposed thigh. As he approached the table, she looked up from the book, and stiffly smiled a welcome.

"Your business over?"

"My business is over."

"Peter says to try again to be friends."

"Fine, I declare a truce. What shall we talk about that's neutral. How about 368?"

She laughed. "Wonderful suggestion. Why in the world do they call it that number anyway? I think 123 is much more charming, or what about 777 for good luck?"

Goodman smiled and sat down beside her. "Whenever a drug company works on a new product, they assign it a number—the brand name comes later and is usually related in some way to its chemical make-up. But at first, it's listed as Frankfurt 368 or FP 368 and after awhile, just 368. You see, no great mystery. Actually, the larger and older companies are up to 5 digit numbers."

"I think three digits are enough for me, thank you. Now let's eat. This place has fabulous spinach salads."

"Sounds fine. What's on your agenda for me after lunch? I promise not be arrogant."

"A walk around this section of the city, which is probably the most interesting. Frankfurt is not like Berlin, or Heidelberg for that matter, but it does have some worthwhile things. My special delight is some of the small photographic galleries. I hope you like photography—you'd better like photography!" She cocked her head to one side in her teasing manner, letting her hair curl over one shoulder while she smiled impishly. "You do like photography, don't you?" she asked, hopefully.

"I do, I actually do," Goodman said truthfully. He had just been to a major show in New York.

The salad was as good as advertised and he enjoyed the meal. But more than the meal, he enjoyed watching her. Whenever her gaze caught his, he blushed self-consciously, but still he couldn't keep his eyes off of her. With lunch over, they began their tour. First through the square, where the town hall stood, the same building that the night clerk at the Hotel Fürstenberg had urged him to visit. Then through narrow streets to the Cathedral of Saint Bartholomew and to the banks of the Main River.

"You get a good view of the city from across the river in Alt-Sachsenhausen," Johanna said, leading him across the historic Iron Bridge. From the other bank, the combination of new skyscrapers and restored historical buildings formed an interesting mixture. But because of the cold wind that came off the river, Goodman found the view less than inspiring.

"It's more pleasant in the summer," Johanna conceded.

"This is definitely indoor weather," Goodman said "museums and that sort of thing."

"We have some nice museums," Johanna said—Museum of History, the Stadel—our art gallery, and, of course, Goethe's house has been turned into a museum."

"I want to do the things a tourist *doesn't* do."

"Then I'll take you back across the river to some of my favorite photographic exhibits. They're off the Braubachstrasse, which is not far."

Hunching up against the wind, she put her arm in his again and huddled close for warmth. Even with the winter clothing in between, he liked the feel of her so near to him. There was something about her that he found very intriguing, yet at the same time comfortable and reassuring. He had to fight off the urge to take her in his arms.

Johanna took him to a handful of small galleries that definitely were not on the main tourists' routes. They were not only tucked into side streets, but they were cramped and with none too good lighting. But the quality of the pictures was excellent, evident even to a novice like Goodman.

"I like realism," Johanna said, as they toured one exhibit of farm photos, "but in painting it seems somehow cheap. Do you know what I mean? In photography, it is natural because it is real at the moment the picture was taken. Do you agree?"

"Not all photographs are candid, though. In fact, often the best ones are posed."

"No, Paul, the best things are always real, never posed." Her tone was earnest. "That is the trouble with people, they pose too much. To be one's self is always best. Uh-oh, there I go lecturing again. Come, I'll show you more."

Outside in the cold again, she huddled close to him for warmth as they peered into one shop window after another. When they stumbled onto Frankfurt's largest toy store, she became excited. "Paul, look at that doll house. It's immense. Just the Christmas gift for my niece. What do you think?"

"I say buy it, it looks great."

"Without even looking at the price? A typical American." As soon as she said it, she tried to retract it. "Sorry, I meant no harm."

"You're forgiven," he said warily.

For the rest of the afternoon they walked hand in hand through the downtown shopping area, stopping briefly in one last photographic gallery. When the evening rush hour began and the crowds grew too dense, Johanna decided it was time to drive home.

Goodman cleared his throat self-consciously. "Peter thought it would be nice if we joined Trudi and him for a night out. Dinner and dancing, 1940s American style."

"Do you want to do it?"

"I'm game if you are. I think Peter needs some cheering up. Bad day at the plant."

Johanna shrugged. "Peter seems very persuasive today. First me and now you. Well, why not. Sure, let's try it."

"But no political talk."

Johanna's face clouded. "Definitely not. You and I would come to blows in two minutes. For Peter's sake, no politics."

"Agreed."

• • •

The nightclub that Werner took his sister and Goodman to, the Nightingale, was in the basement of one of the city's plusher hotels. A separate entrance at street level permitted easy access, without the need to enter the hotel proper. The doorman recognized Werner and greeted him warmly, but this was nothing compared to the effusive welcome from the headwaiter, Becker, a small man whose missing left arm did not prevent him from embracing Werner. Becker led them to a choice table at the side of the dance floor, cozily tucked into a recessed wall. The decor of the room was muted, overhead lights radiating soft, pastel colors, plain white tablecloths bathed by gas lamps located at each table. The room was almost three-quarters filled and a steady hum of conversation was emanating from the patrons, most of whom were dining as well as drinking. A large curtain covered the rear of the dance floor.

Cocktails were ordered when a waiter appeared. Johanna seemed bored, her sister tired. "Where's the famous music I've heard so much about?" Johanna asked.

"Be patient," Werner said, laughing. "We have a little time before the first show; then you'll see what all the excitement is about."

"That assumes there will be some excitement," his wife interjected. "After taking care of those children all day, I'm more suited to sleeping than dancing."

"I thought it was an opportunity to get you away from them for a little while," Johanna said. "And a chance for you, Peter, to forget all those problems at Frankfurt Pharmaceutica. I think you've aged five years since your last birthday."

Werner grimaced. "Sometimes I think you're right. This 368 project is definitely causing my hair to turn gray, or fall out, or both. But I'm glad we're here. I was so depressed at hearing about Hawthorne this morning, it ruined my whole day."

"Who's Hawthorne?" Johanna asked as the waiter returned with the drinks.

78

Werner explained the ill fortune that had befallen the English cardiologist and its implications for the drug study.

Trudi shook her head in sympathy. "That poor man. I'll bet it was because he was working on that stupid drug."

"Don't be ridiculous," her husband snorted. "He was robbed like people are robbed every day. What in the world did 368 have to do with it?"

"It's a jinx, if you ask me. Call it woman's intuition."

"I call it poppycock," Werner snapped back at her and the two of them glared at each other.

"If that isn't enough," Werner continued, "I've had one of my most valuable foreman, Meyer, take off on a skiing holiday. There's something peculiar about him. He flirts with the neo-Nazi movement and yet he has Richter's ear. Richter is always calling them 'demented.' So why protect one of them? Worst of all, Meyer seems to be the only who who knows where the supplies for making 368 are kept. We can't make any more for the research studies until he returns."

"Where did he go?" Trudi asked.

"To the Harz Mountains. I left messages at his hotel to call me, but he hasn't done it. I think he's avoiding me."

Just then, the curtain near the dance floor opened, revealing a bandstand. The audience applauded as the twelve-piece band swung into its first piece, "One O'clock Jump." The female singer came out, sequined dress slit to the thigh, spiked heals, hair swept up high on her head, her mouth bright with red lipstick. In a high-pitched voice, she sang sometimes in German, sometimes in English, but always in the clipped tones of a singer following music scored for a band rather than the other way around. She began with a medley of Glenn Miller tunes—"Perfidia" was her best effort—then stepped back as the band took over.

The dance floor became crowded. "It's now or never," Werner said to the others, leading his wife toward the music. "Soon there won't be much room." Holding Johanna stiffly in his arms, Goodman led her through a series of ballads from the golden age of swing. The more they danced, the more she relaxed.

"I like the music," she murmured. "Even if it is American." She stuck out her tongue at Goodman, who said nothing.

"This was a perfect place to come tonight," she told her brother-in-law when they returned to the table. Even Trudi seemed pleased, her weariness forgotten. They all sat for awhile, listening to the band work first on a medley of Tommy Dorsey tunes, then one from Benny Goodman. When it was time to go, the headwaiter was effusive in his good-byes.

"Why is he so attentive?" Goodman asked, as they were getting ready to leave.

"Becker used to work for Frankfurt Pharmaceutica," Werner explained. "He was actually quite a good chemist. He helped my father with some of my lab work. When he decided to do this instead, I gave him excellent references. And he's done well here. Not only is he the headwaiter, but he's a partner now, or something like that. I suppose he feels my help was instrumental in getting him started."

"Peter's too modest," Trudi added, "he also gave him a little loan when this place looked like it might fail. I warned him against it—I never could understand its appeal—but he turned out to be a better judge than I was."

Werner smiled. "That's about the first time I've heard her admit such a possibility! But I really felt this place was worth supporting. And anyway, Heinz had been very good to my father for many years when his health was failing. It was the least I could do."

"But your father was good to him, too," Trudi protested, "putting up with his slowness and all, because of his arm."

"A war wound," Werner said. "I suppose my father was more understanding because of that." He paused. "But I've a busy day tomorrow—one of those dreadful meetings with Stern—so let's get home. We'll drop you on the way, Paul. By the way, why don't you and Johanna take a little Nordic skiing trip yourself? Maybe go over to Herzberg and ask Hans Meyer some questions for me."

"Are you joking?" Johanna asked.

"Why not? You need a rest from those books you're always reading. And Paul needs one too. Do your brother-in-law a favor and take a holiday at the same time." Werner put his arm around Johanna. "How can you refuse such a request?" he said with mock seriousness.

Johanna was not about to give in that easily. "Don't you have to be back in New York, Dr. Goodman?"

"Nope, wouldn't mind a few days off at all. Of course, I don't know how to do Nordic skiing and I don't have any clothes . . . "

"No problem, both matters can easily be dealt with. You see Johanna, you're the obstruction, not Paul. Paul's willing."

"But Peter . . . " Johanna stopped in mid sentence when her eye caught Trudi's glare. "Alright, I'll do it," she said without enthusiasm. "One last favor for you, then I'm going back to the university."

"The last favor for *now*," Werner said, laughing.

Frankfurt, January 7

It was nearly noon by the time Goodman checked out of the hotel. He was assisted by the same clerk that had been on duty the day he arrived in Frankfurt.

"Was your stay a pleasant one?" the clerk inquired as he pulled out the ledger to see if there were any incidental charges that the guest—and not Frankfurt Pharmaceutica—was responsible for.

"Excellent. Now I'm off for a little skiing."

"Where to, the Alps?"

"No, just the Harz, near Herzberg. Cross-country skiing we call it. It's what you call Nordic."

The bald head glistened. "I wish I were there myself," he said sighing. "Lovely spot."

Johanna was waiting for him in her bright red Porsche. In addition to bringing several outfits of her own, she had borrowed her brother-in-law's ski clothes. Goodman stuffed them into his suitcase and placed the bulging bag in the back seat. "Do you think Peter's things will fit me?"

"Why not? They should. You're about the same size."

"And the boots and skis?"

"You can rent all of that. My equipment is back in Heidelberg, so I'm going to be a renter as well. Climb in and let's get started; it's a long drive—over 250 kilometers. I can't believe the things I do to keep in my sister's good graces."

"Why don't you let me do the driving?"

"Maybe later. First, I have to work out my aggression on the gas pedal. This is definitely the *last* favor I'm doing for either Peter or Trudi." Having said that, she seemed to calm down. "I'll get us to the autobahn and drive at least the first half of the trip, then

82

we'll see about you taking over. Now get into the car so I can show you the route on the map, I'm very compulsive about keeping my navigator well informed."

"Yes, ma'am," said Goodman, doing as he was told.

Johanna spread a map across his lap and with a pencil, pointed out the roads they would take. "We'll leave Frankfurt via national highway 3 and then take the Frankfurt-Kassel autobahn. After we pass Kassel, we'll come to Göttingen, where we'll leave the autobahn system again and take a smaller road east to Herzberg right at the foot of the Harz mountains. After we talk to Meyer, we might as well do some skiing. We can either stay in Herzberg or drive around the shores of the Oderstausse, a really gorgeous lake; we can stop at inns in any of the villages nearby." She pointed to several dots on the rim of the lake. "They'll all be accessible to good ski routing trails. Any questions?"

"An excellent geography lesson. My compliments, Miss Bauer." She gunned the motor, effectively ending any further conversation for the moment. The trip had begun. Neither Frankfurt's narrow winding streets nor crowded boulevards presented problems for her. Skillfully maneuvering the Porsche through traffic, she soon was out of the heart of the city and onto the long-distance highway leading to the Kassel autobahn. The day was clear and she hummed along with the music from the car radio.

"Isn't it a gorgeous day?" she exclaimed as they raced north. "There's a report of snow showers in the Harz region, but nothing around here. I wonder why Meyer went so far away when he could have gone to the Taunus, a much closer range."

"We'll ask him when we find him."

"Hopefully, it won't take long. Aren't they going to need you in New York?"

"They don't expect me back until Monday. I was planning to take a long weekend after this trip."

"And your patients, what about them? Who looks after them? You do have patients, don't you? Or are you a research doctor, like Peter?"

"No, I'm a *real* doctor. I see patients and I treat them. But the hospital is the center of my activities—that's where I have my office and laboratory, that's where I teach my medical students."

"You sound very possessive. All those 'mys.' "

"I didn't mean to imply . . . "

"Typical male style. Unfortunately, just as prevalent here as in America."

Goodman laughed. He knew if he didn't take her too seriously, he could enjoy her company. Settling deeper into his seat, he alternated between watching her and the open countryside. He decided he preferred her.

"You're staring at me too much," she said. "Remember this trip was Peter's idea, not mine."

"I don't think Trudi liked the idea of our going off together any more than you did. Am I right?"

"Trudi knows me well enough to know I can take care of myself," she said firmly.

"That I'm convinced of. Have you ever come close to marriage? With your intelligence and good looks, I'm sure you've had ample opportunities."

"I almost did a few years ago when I was twenty-four. I was living with my history professor. Gerhard had divorced his wife and changed his lifestyle to marry me, or so he said. You know the type: the respectable fellow in his mid-forties who wants to be thirty again, dresses like a teenager, buys a sportscar, the whole works. After a while, I figured out that the change was not for my sake, but for his, a part of his mid-life fantasy. So I moved out. I can be rather

84

independent-minded at times. And I especially don't like to be trifled with."

"Yes, I've noticed that."

After two hours on the road, they stopped for a brief lunch, then resumed the journey. Traffic thinned out; driving was pleasant. In the late afternoon light snow began to fall, and by the time they reached Göttingen the surrounding countryside was all white. The Harz stretched out before them.

Goodman had taken over the wheel; Johanna napped in the seat beside him. The driving was slower with snow on the ground, and it took almost an hour to reach Herzberg. Though the area they were passing through was relatively flat, the horizon was completely dominated by the Harz. To Goodman they seemed a vision of winter loveliness. Not overpowering like the Rockies or the Alps, but serene and majestic just the same. The fading light of day, the falling snow, the complex and enigmatic woman dozing next to him—he was aware of it all blending together into a feeling of adventure that he could not remember experiencing before.

When he reached the outskirts of Herzberg, she awoke. The winter sun was setting early. "It's getting too dark to see the lake," she decided, "we might as well find the inn where Meyer is supposed to be."

"One quaint little inn, coming up." It actually didn't take him long to find the one that Peter had told them about. According to the guest registry, Hans Meyer was still there, but no one answered the phone in his room. "We'll have to find him later in the dining room," Johanna decided, "or on the ski slopes tomorrow." They checked into separate rooms and unpacked. An hour later, as planned, they met in the dining room. Goodman had already surveyed the inn—no signs of Meyer anywhere.

"We might as well enjoy dinner," he said to Johanna. "I doubt we'll find him tonight. We'll have to do it tomorrow, before he takes off and goes God-knows-where."

Goodman found the dinner of homestyle lamb stew to be excellent, and the conversation surprisingly pleasant. Johanna seemed less hostile, more relaxed. Was it possible, he wondered, that she was actually getting to like him? With that comforting thought dancing around in his head, he was almost tempted to kiss her goodnight when they parted later in the evening. But she moved too quickly and the opportunity was lost.

London, January 8

London was cold and damp. As soon as he left the heated Heathrow terminal, Tamir had trouble adjusting to the uncomfortable weather, despite his warm raincoat. He had not planned to be in England, but when his colleagues at the university could offer no help with the cyanide-like formula he showed them, he told Shlomo that there was only one person he could turn to.

"Professor Eugene Lester of the University of London. He taught me everything I know. He's visited Israel; he likes us. In other words, he's trustworthy."

"I was going to send you directly to Frankfurt," Shlomo said, "I'm under a lot of pressure from the Prime Minister's office."

"Let me call Lester. If I can see him, it's worth a stopover in London. He has close ties to the British chemical warfare people. He's a goldmine of information. What do we have to lose?"

Reluctantly, Shlomo acquiesced. Tamir called Lester the next morning and arranged a meeting for the afternoon of the following day. Shlomo had Tamir put on the January 8th El Al morning flight to London in one of the first-class seats reserved for government couriers. "Stay at the Eustace Hotel; it's near our embassy. One of my people will contact you. And bring a heavy raincoat."

Tamir did as he was told. By 2 P.M. he had checked into the Eustace Hotel on Knightsbridge Road. With the assistance of the concierge—and a map of the London underground system—he was enroute to his appointment several minutes later.

The few times that he had been in London, he enjoyed the underground, the movement of large masses of people who seemed courteous and friendly, the babble of many different languages. He found the Notting Hill Gate station without difficulty, and four stops later ended up at the East Acton station. From there it was a 15-minute walk to the Royal Postgraduate Medical School, where Lester had his office. The area had once been populated by Englishmen, but immigrants from the West Indies, India and Pakistan had changed that completely. Despite roadside graffiti and a slightly run-down appearance, the two-story houses seemed well kept. "Down With Western Imperialism!" threatened fiery red letters on one playground wall, but otherwise the neighborhood had a benign air to it. Even the dark and grimy pedestrian subway seemed safe. Tamir walked on through what he thought were clearing skies, but just as he finished his walk, the skies darkened again. Fortunately, when the rain started he had already entered the medical school building. Lester greeted him warmly.

"Good to see you again, Dan. How long has it been? Three years?"

"It was four years ago that you were a visiting professor in Tel Aviv. And you don't look a day older."

Lester's oval face beamed in delight. "You don't say. Not a day older. I just turned 60 years today, so I'm very pleased at that bit of flattery but I don't know if I'm up to snorkeling at Eilat again. That was quite a lot of strenuous exercise you had me doing."

"But we had a wonderful time."

"We did, we truly did. Here, sit down and tell me what's on your mind. As I recall from your phone call, it's a bit hush-hush."

"The official word is 'discretion advised.' "

"In other words, no babbling after two Scotches."

"Exactly."

"Alright, ground rules accepted."

Tamir showed him the piece of paper. "It's the formula for the cyanide derivative I'm interested in. I'm concerned about airborne spread."

Lester studied the sheet of paper for several minutes. "You're right, of course," he said finally, "this particular derivative can probably be made into an aerosol. Nasty business. Where'd you get this? The writing looks Arabic. And what's this other formula?"

"Can't say where I got it and I'm not sure what the other formula represents."

"I didn't mean to meddle, but I assume this material is from an unfriendly source."

"You've assumed correctly."

"Too bad. Yes, I've seen this type of structure before. You've come to the right place. Actually, some of our people have been playing with this problem for awhile. I can let you look at the files on it if you like. Would you care to see them? *Your* discretion is needed now, I'm afraid."

"No problem."

"Good. I'll have my secretary get them while we have some tea." He looked at his watch. "But first, I have an errand to run. Friend of mine was injured yesterday and is in the hospital wing across the way. I promised him I'd pop in and say hello. It will just be a minute. Come along, the walk will do you good." Lester gave instructions to his secretary and then led Tamir through a winding corridor to the hospital. His colleague, an older man, was propped up in bed in an open ward, his right arm in a cast. Lester introduced Tamir, then asked solicitously, "Hawthorne, my dear chap, how are you today?"

"Tolerable," Hawthorne grunted. "But difficult to do any work like this. That's my writing arm they got. Imagine being beaten leaving the underground on the way to work. Never heard of such goings on before, even in this neighborhood. What's the world coming to?''

"Surely you can let the work accumulate for awhile. This is no place to do anything.'' He gestured around at the crowded ward that Hawthorne was in with beds stacked in rows in the traditional British manner.

"I feel I must attend to my mail at least, even if I'm not up to carrying on with my other work. Ah, here comes my secretary now with today's load.''

A pleasant but rather plain looking woman deposited a pile of letters and scientific journals on Hawthorne's bed. There was something about her features that puzzled Tamir. He had seen her face before, but couldn't quite put his finger on where. After chatting a few minutes with Hawthorne, Tamir and Lester excused themselves, leaving Hawthorne buried in his mail.

"Is Hawthorne a Ph.D. also?'' Tamir asked.

"No, an M.D., but he does a lot of research. He's a cardiologist. Darn good chap.''

"His secretary with him a long time?''

"Elizabeth? For years. A real work horse. But why not? She's unmarried and from the looks of her, she'll stay that way. Good old Beth.''

Walking back to Lester's office for afternoon tea, Tamir recalled where he had seen Elizabeth's face before: in a photograph taken in front of Harrods. She had been with her "boyfriend,'' the since deceased Libyan army officer. "Good old Beth,'' Lester had said, that must be where the "B'' on the letter had come from. Shlomo would be pleased with this bit of unsuspected news, but he would also want him to find out more about Miss "B.'' Tamir wondered how he was going to do that.

Herzberg, January 8

Friday in Herzberg was glorious. Bright sunshine streamed into the room while outside birds chirped in a rich cacophony. Yawning and stretching, Goodman surveyed the scene below. A comfortable layer of snow lay over the valley and the gentle slopes of the surrounding hills. Figures on skis were already dotting the trails. Dressed warmly in Peter Werner's sweater and knickers, he added as a final touch a down vest and multi-colored ski hat. Johanna was waiting for him in the lobby. Meyer had not checked out, but the stocky blond youth was again nowhere to be seen. They would try the ski areas. With chocolate bars packed into their pockets, they cleaned off the Porsche and followed a van to the ski touring area. In a large cabin with a great open fireplace, they rented skis and poles. Hot coffee and chocolate were served in earthen mugs by young girls in billowing peasant skirts. Equipment in hand, the dozen or so people in the cabin gradually moved outside again, fastened their skis, flexed their poles and milled about chatting in groups of three or four. Goodman was able to fasten his skis to his boots without difficulty, but when he tried to stand up he fell sideways, landing not-too-gently on his shoulder. Johanna stifled a laugh, but none of the others paid attention. They had all been novices at one time or another. Goodman wiped the snow from his face and, following Johanna's directions, lifted himself by shifting his weight while bending his knees in a series of jerky maneuvers. With a determined expression on his face, he stood upright once more. But the feeling of pride was not too long-lived, for as soon as he tried to move forward, he fell again. The whole process of standing up had to be repeated. After several more attempts, he was finally able to propel himself forward without constantly winding up on his back or belly. Much to his relief, the others in the group avoided any comments.

"Don't be discouraged, you're doing fine," Johanna said, "but try not to be so stiff. Stop trying to lift the ski and take steps with it. You're *not* supposed to walk, you know, you're supposed to *glide*."

"Sure, it's easy for you to say," he responded from his precarious perch. "You were born on skis."

She laughed. "Not quite."

He chose that moment to fall again, and again she held her glove to her mouth to stifle a laugh. When he finally righted himself, she told him to try balancing himself better—he was tilted too much to one side. "And use your poles correctly. You tend to lose your rhythm with your foot movements. Here, it's like this—arm forward, leg back. Do it the way I'm doing it." She proceeded to show him. "Remember, you have to learn to get that rhythm, that cadence."

He listened, he learned. He watched her, he did it himself. Gradually, he began to master the gliding movement, the rhythm of the arms and the legs, the use of the poles. "Hey, I like it," he called over his shoulder to her, as they skied single file down a trail made by the skis of the several groups ahead of them.

"I told you you would," she yelled back.

They spent the better part of the next two hours skiing, finally stopping for a rest when Paul complained he was out of breath.

Grunting with relief, he sat beside her on a large rock. "Lord, that winded me," he said while they unstrapped their skis.

The snow had stopped falling, the air was fresh and clean. "Look down there," she said pointing through trees. "That's Herzberg. See the church steeples, and the smoke coming out of the houses? Lovely, isn't it?"

He had to admit it was. A pretty countryside. "Must be great in the autumn as well, with the leaves turning."

"Beautiful. In fact, it's a four season type of place to get away to. I like it. Peter and Trudi and Gerhard and I"—she stopped abruptly.

"Hell, you can talk about him. I don't mind."

"It wasn't Gerhard that stopped me. I was thinking of Peter and Trudi again. He seems so troubled lately. It's taken its toll on his marriage, that's why it was good to see them out dancing. Sometimes, I think Trudi's right; 368 is a jinx. But we can forget that for awhile, can't we? In fact, it's more fun without Hans Meyer to think about."

He looked up at the hillside. "Ooh, it's so steep. I can't even see the top. What's wrong with stopping at this point?"

She shook her head. "You have the athletic ability of a snail. Gerhard would be bounding up into those thickets like a hunting dog. And he's ten years older than you."

"I'm not Gerhard. I'm sorry."

Talking about her ex-boyfriend had bothered him. "Let's start down."

"Sure," she said, quickly. "It will be easier going down."

"If I don't hit any trees at fifty miles an hour."

She laughed with him. "You won't, just follow me."

Down she went, turning and twisting, constantly braking with her skis, not allowing too much momentum to build up in the heavily wooded areas. Halfway down, she stopped and waited for him. He had trouble with several trees, but nothing he couldn't handle. "I'm doing okay," he shouted. "I want a suitable reward."

"A cup of hot cocoa in the ski cabin?"

"Sounds good," he shouted back at her.

"Okay, but first make it to the bottom in one piece. Remember, go slowly. The tracks are not that good coming down."

Using his poles the way she had taught him, he followed her down. Occasionally, they would break into an open stretch where he could almost pretend he was a downhill racer, but most of the time it was stop and start, dodging bushes and trees as they went. At the bottom of the hill, she took off across the field, moving quickly with the grace and rhythm of a born athlete. He admired her form as he skied behind her. Over the little bridge, onto the other side, the tracks were deeper now and they whistled across the field. A cloud or two had appeared to mar the otherwise perfect blue sky, but it still was a gorgeous afternoon. They moved across the field to the same cadence. She had taught him well.

The ski cabin was crowded. It was early afternoon and people were in every corner eating sandwiches or cheese. They had to wait several minutes before finding an open place to drink their hot cocoa.

"Where did all these people come from?" Goodman said in wonderment.

"It's a popular area. They arrive at all hours."

The cocoa was burning hot and they sipped it very slowly. "You did well for a beginner," she said, complimenting him.

"But I'm still no Gerhard." His tone was light and she responded in kind. "You have some good points," she said. The way she looked at him surprised him. There was actually warmth in her eyes.

The rest of the skiing went much better for him. He was more confident now and used the poles with more assurance. Spills were rare. Frolicking in the snow with her was an unbelievable delight. Only when the sun began to fade did they call a halt. They raced back to the cabin. He went faster than he thought possible, but she beat him easily. While she returned the skis and poles, he cleaned the snow off the front and back windshields of the Porsche. He felt exhilarated; the day had been perfect. Quickly, he constructed a heavy, wet snowball and hurled it at Johanna as she came out of the cabin.

"Stop that," she yelled. The snowball had struck her on the shoulder. "Sorry," he yelled back.

"It hurt," she snapped in mock anger and threw some snow at him.

"Missed." He was already collecting snow for the next missile.

This next ball was even larger than the first, but she was ready for him now. As he released the snowball, she ducked and it flew harmlessly over her head, striking the hood of a black Citroen that was passing behind her. The driver slammed on the brakes and peered out angrily at them. Goodman spread out his hands, palms raised. "Unintentional," he said. "Sorry." The driver said something unintelligible under his breath and drove off. Giggling, they watched him leave the lot. Still eying the car as it exited, Goodman's gaze paused at a familiar-looking figure. Squinting against the sun, he said to Johanna, "Doesn't that guy look familiar?"

She looked where he was pointing. A man with a short well-trimmed beard was just getting into an Audi. "Yes, I think I saw him in the restaurant where we had our cocoa."

"There's something about him that bothers me, but I don't know what. He must work here; he certainly isn't dressed for skiing."

"I shall call him 'our mystery man,'" Johanna said, laughing.

They unloaded their gear into the Porsche and Johanna drove cautiously back through the town. More pedestrians were about and they spilled out into the thoroughfares. The inn was a welcome sight, brightly lit, warm and comfortable inside. Half a dozen people in their early twenties were grouped around the piano in the music room, singing folk songs and lending a festive air to the wintery dusk. Goodman nonchalantly put his arm around Johanna and they joined in the singing. "Well, well," he said wonderingly, when the last note had been plunked. "Look who's staying at our hotel."

"Who?"

"Our bearded friend."

The man was at the front desk, picking up his room key. "Do I know him or don't I? Do me a favor, make friends with the desk clerk and find out who the bearded fellow is."

"How do I do that?"

He grinned. "Oldest trick in the world. Just bat those baby blues at him, make small talk, and after a minute or two he'll tell you anything you want to know, including how much money he has in the bank. A pretty woman can get anything she wants out of a man with a little sweet talk."

She looked dubious.

"Try it," he insisted.

"All right, I'll try it. You sing along with these nice people while I try my charms."

She ambled over to the desk and rested an elbow on the registry, a big smile on her striking

96

features. Goodman remained at the piano and joined in somewhat out of tune, but robust, when the group singing resumed. Presently, she returned, a puzzled look on her face.

"Well," he said, "I'm dying with curiosity."

"He's some sort of Romanian engineer on a business trip for the Romanian state water works, Anton Ionescu, or something like that."

"Thanks for the reassurance. There's no way on earth I'd know somebody like that. Obviously a case of mistaken identity. But *that* person I know," he said, indicating a new arrival in the lobby. There was no mistaking the blond good looks and full moustache of the plant foreman. Hans Meyer had finally surfaced.

"There's our man," he said to Johanna. "Might as well get this over with now."

Meyer looked up in surprise when Goodman touched his arm and said, "Peter Werner sent me to . . ." but Goodman never finished. Meyer bolted for the door. Goodman grabbed him by the arm. "Hey, wait a minute."

"I have nothing to say to you," Meyer shouted. "I'm on holiday, leave me alone."

"Dr. Werner needs the 368 supplies. Only you know where they are."

"368? Is that what he's bothered about?" Meyer seemed relieved.

"Sure, what else would he be concerned about?"

Meyer looked at him skeptically. "I don't believe you. He didn't send you all the way out here because of that."

"Why not? Why are you so damn suspicious?"

"That's none of your business," Meyer shouted and brushed past Goodman. When Goodman grabbed his shoulder again, the husky youth swung around quickly and punched him twice in the abdomen. Goodman gasped in pain and fell to his knees. While Goodman writhed on the ground, Meyer walked quickly away.

"I'm okay," Goodman insisted, as Johanna hovered over him protectively. "Just had the wind knocked out of me."

"It was a sneak punch," Johanna said angrily. "You never had a chance to defend yourself. What could have made him do that? What a strange man. Wait until Peter hears about this."

Slowly Goodman walked to his room, with Johanna next to him. He stretched out on his bed and within a few minutes felt better.

Johanna sat on the edge of the bed, one leg crossed over the other and eyed him sympathetically. "I'm sorry you were hurt, even if you are a very disruptive influence in my life," she smiled. "And to think, I was planning to spend my visit with Trudi studying. How boring that would have been."

"I'm glad you decided not to study too much," Goodman said. On a sudden impulse, he reached up and drew her to him. She didn't resist. Their kiss was long and passionate. While she sat beside him caressing his face, he slowly unbuttoned her sweater. "I'll freeze," she protested half-heartedly, but as the sweater fell away exposing her brassiere, he embraced her again and she yielded easily in his arms. He undid the snap in front of her brassiere and the two soft, full breasts fell into his hands. He quickly took off his own shirt and drew her tightly against his bare chest. In a few seconds, they both were completely naked, their hands and mouths all over one another. His passion fully aroused, he pushed inside her, thrusting deeply, feeling himself rising and falling with her as she clung to his neck and moaned in delight.

Her moans were louder and more frequent now, her mouth open and sensuous. They rocked back and forth until both reached climax. Slowly, the spasm ended, and they lay together, still not separated. He kissed her, a long full kiss, and then finally withdrew from inside her.

"Look at the mess you've made of this room," she said laughing. "You've thrown our clothes in all directions." Then she grew more serious. "I don't know why I'm laughing. This is one situation I never expected us to be in. And with someone who'll be five thousand miles away in a few days. Now what do we do?"

"Being with you is one of the best things that's happened to me in a long time. Why should I rush back to New York? I'll make this into a real vacation and spend it with you. That is, if you want me to."

"Of course, I do, you idiot," she was smiling again.

"I'm afraid I'm not much of a skier."

"You're getting better." He reached out to her hesitantly. But she was happy again and she came to him eagerly, her body warm and ripe, her lips seeking his.

They spent the rest of the afternoon in bed making love and by evening they were both exhausted. Even the arrival of the bearded Romanian engineer at the table next to theirs in the dining room did not bother them. By mutual consent, they dismissed both the Romanian and Meyer from their conversation. There would be time enough later with Peter to discuss the blond pugilist's violent reaction. They slept late on Sunday—and by the time they took to the trails, they were almost obliterated by the throng of weekend skiers.

"It's awful," Johanna protested, "why don't they go away and leave the place to us?"

"Wait a day, they will. Now's our chance to take a drive around the lake."

The Oderstausse was not frozen and the cold clear waters formed a stunning blue contrast to the white mountains with their patches of green. Goodman felt a sense of contentment that he wished could go on forever, a fantasy that he realized was pure wishful thinking.

When they returned to the hotel in the late afternoon, a note was waiting in the key box. "Please call the local police station—66421." When a puzzled Goodman finally got through to the officer-in-charge, the message was terse. "We have been looking for you and Miss Bauer for several hours, Doctor Goodman. There has been a serious traffic accident in Frankfurt involving Peter Werner." Goodman's body felt like he had been struck by Hans Meyer's fist all over again.

"What hospital is he in?"

The officer paused. "He was taken to Central Hospital yesterday. His wife asked the Frankfurt police to contact the resorts in the Harz region. But I'm afraid he is no longer in the hospital. He died early today."

Stunned, Goodman placed the receiver back in its cradle. The lake and the mountains seemed as far away as the moon.

PART 2

The Death of Peter Werner

Frankfurt, January 9

It was dark when Goodman and Johanna arrived back in Frankfurt. He had driven the first half, she the second, but the trip had been difficult. Traffic was often heavy and with each delay, Johanna would begin sobbing. As much as she was upset by the death of her brother-in-law, she was also very concerned about the emotional well-being of her sister, now left alone with two small children.

There seemed little that Goodman could say that comforted her and, indeed, after the initial outpouring of grief when she had clung to him, he felt a distance growing between the two of them. Except for the times that traffic brought them to a halt and she cried, the interior of the Porsche was unnaturally quiet as they retraced their route from Herzberg to Frankfurt. Goodman realized that a large part of the sudden estrangement was the guilt that Johanna felt. While she played in the snow—or in bed—her brother-in-law lay dying, her sister was in a state of shock, and her niece and nephew were confused and afraid. It was no wonder that she was angry at herself.

When they finally reached the Werner house, Johanna pulled the Porsche sharply into the driveway and ran into the house, Goodman following slowly behind. By the time he entered the living room, the two sisters were already in each other's arms. Their tears ran together as they embraced one another.

"How did it happen?" Johanna asked, sobbing.

"A truck smashed into him on the highway from Frankfurt," her sister said, fighting to regain her composure. "The police were there almost at once, but there was little they could do. When they brought him to the hospital, he was nearly dead, in a coma. He never woke up, never saw me, never talked to me—it was unbelievable. Now I've resigned myself to it, I have to for the children. They've been wonderful. I just put them to bed, trying hard to smile, not to cry, but it's so difficult. It will be bad again at the funeral."

"When will that be?" Johanna asked softly.

"On Tuesday, the thirteenth."

Goodman came forward to extend his condolences.

"Thank you," she said, now fully composed. "I know how much Peter valued your friendship. It was something very new and different for him. He was looking forward to seeing you again in the future."

"If there is anything I can do . . . "

"There is, Dr. Goodman." She paced back and forth, stopping only to emphasize a point. "After the initial shock of Peter's death wore off, I began to think back on the last several months. My husband was not his usual self, Dr. Goodman, and for reasons that I can't articulate well, I don't know if his death was truly accidental."

Goodman and Johanna exchanged puzzled glances. "Why do you think that?" Goodman asked.

"As I said, Peter had not been acting like himself for awhile now. Even Johanna had noticed it and she doesn't really get to see him that much. But Peter wouldn't say what it was that was bothering him. In fact, he denied there was any problem at first. I thought I had said something or done something to upset him, but he said no, everything was all right. Then I thought he had another woman."

"Ridiculous," Johanna said.

"Yes, finally that is what I said, too. And after awhile, I stopped thinking such thoughts. I decided there could be only one thing that could be making him so strange, so depressed. His work. And you know, Dr. Goodman, the more I watched him for little signs that I was right, the more I saw."

"What do you mean?" he asked.

"When he first came home, he would be in a foul mood, not even the children could cheer him up, and you know how uncharacteristic that is of Peter. Gradually, he would feel a little better, but by the next day, just the thought of going to the plant seemed to upset him. One day I asked him outright, 'What is going on at that stupid plant that leaves you so unhappy?' He seemed surprised that I had guessed his little secret. He shook his head in disgust. Better not to bother you with it, he said, and not a word to the children. So then, of course, I knew I had guessed correctly—it was something with work. Well, what's the biggest thing in his life at that place?"

"The new drug 368?"

"Exactly."

"There must have been something more," Goodman insisted.

"I think he was frightened, too."

"Of what?"

"I don't know. All I know is that one night I saw him with a pistol. He seemed upset that I had seen him. I think it was his father's pistol."

"But you don't know he was using it to protect himself."

"No, I don't. But he had never had it out before."

No one said anything for several moments.

"How can Paul be of help?" Johanna asked finally.

"I don't think Peter's death was really an accident," Trudi said, trying hard not to cry, "but no one listens to me, especially the police. Maybe they would listen to Dr. Goodman, a distinguished visitor, a physician. I know you're supposed to be going back to New York, but could you stay longer? I'm not one to ask favors of people, but I just feel you could accomplish more than any of us."

Goodman looked at Johanna. "Of course I can stay. I'm flattered that you would ask me to help."

"We all could use some rest," Johanna said wearily, "especially you, Trudi. It's late and it must have been a grueling day for you."

Her sister nodded her agreement. "Dr. Goodman, why don't you stay here in our guest room tonight. It's easier than trying to find you a hotel room now."

"Thank you."

After Trudi had retired, Johanna took him into the kitchen to make sandwiches for a make-shift dinner.

"I don't know whether Trudi's imagining things or not," Johanna said, "or whether it's just a case of emotional shock."

"You still want me to stay on awhile, don't you? I mean, you have no second thoughts about yesterday in Herzberg."

Johanna put down her sandwich. "I don't know whether you can ease Trudi's concerns, but yes, I would find it very comforting if you were here. At a time like this, it's good to have someone that you feel strongly about close by." She bent over and kissed him. "And don't worry, I have no misgivings about yesterday."

In bed later, Goodman pondered the day's events and their implications for him. The death of Peter Werner had been a severe jolt, but to hear Werner's wife hint at possible foul play was even more of a jolt. Why would anyone wish Werner ill? But then there was that business with Meyer. What was Goodman's responsibility in uncovering the truth? If it had not been for Johanna, he knew he would feel less inclined to stay and help Trudi. But that, of course, was the problem: Johanna. He had formed an emotional attachment to her, one that had grown deeper than he ever thought would be the case when they went on their skiing jaunt.

This woman had touched something within him; feelings that had lain dormant for years were now stirring again. He had even played with the idea of having her visit New York in the spring at his expense to see, indeed, if this could really be more than just a short-lived fling. Now he had the opportunity to be of some help to her and her family and at the same time see her in a totally different setting. What better way to evaluate the depth of his feelings for her? It would not take a great effort to extend his stay—a call to his department head at the hospital to make arrangements for coverage of his clinic by an associate. His research laboratory could easily be run by his chief technician for another week or so. This was done anyway during his regularly scheduled vacation period. He could not expect Frankfurt Pharmaceutica to pick up further hotel bills now that his regular visit was concluded, but staying at the Werner home would take care of that problem. What actually could be accomplished, he didn't know, but if asking a few questions would help soothe Trudi's anxieties, it would be the least he could do for the kindnesses the Werner family had shown him. So resolved, he fell asleep.

At breakfast with the Werner family, he spoke of Trudi's request. "I'll do the best I can, but I'm afraid it may not answer all your questions."

"Whatever you do will be greatly appreciated," Trudi said.

To the children, Goodman said simply, "Your father was a very fine man. I'm going to stay with you for a few days to help make things easier for your mother, but if you need me to help you with something, I can do that too. All right?"

Both heads nodded, but how much they understood he did not know. The enormity of their loss was probably still too much for them to comprehend. After they had gone upstairs, Goodman told the sisters of his plan. "The first place to start is with the police. It would be best if one of you would come with me."

Trudi agreed. "Johanna, could you? I'm not sure I could do that again."

"Of course, Trudi. I'll help you clear the table, then Paul and I will go."

"I'll have to call New York and tell them I'll be delayed," Goodman said.

"Do it now," Trudi insisted. "Use the phone in her living room, it's quieter there."

As soon as he had arranged his schedule, he and Johanna climbed into her Porsche and drove downtown. The Frankfurt police station that had handled the accident was not far and the officer in charge of the case, Inspector Slitko, was young, crew-cut, and courteous. He took them to a huge, unheated garage behind the station house. On either side, battered hulks were neatly laid out. In the middle of one section was Werner's brown Volkswagen.

The inspector leaned toward Goodman. "As you can see, Doctor, it was a miracle he lived as long as he did. There was really nothing the hospital could do. It took the emergency unit several hours just to pry him out." Even though his voice was purposefully low, Johanna heard and shuddered.

"I know you've been through this before, Inspector," Goodman replied, "but if you would, could you reconstruct it for us?"

"I don't mind at all," the policeman said sympathetically. "I understand what the family is going through. Here, come around to this side and let me show you some things. And you too, Miss Bauer."

They walked around the crumpled chassis while he pointed out where the huge trailer truck had struck the driver's side of the car and run right over it. The door was wedged into the engine. Dried blood was everywhere. "A mess, you see, a horrible mess."

"There were witnesses?" Goodman asked.

"Yes, several. All tell the same story. The driver behind Doctor Werner saw the whole thing. He stopped, tried to help, gave us a complete version. It was twilight, you understand, a bad time for that highway, full of commuters. A slight mistake at those high speeds . . . In any event, Werner's car was in the right lane ahead of this witness' car and apparently not going very fast at all when the car started to show erratic behavior. It swerved into the center lane rather suddenly, and was struck by the truck. The impact of the truck on the side of the Volkswagen distorted the chassis and sent it spinning. Right off the road and down the embankment." He clasped his hands behind his back and stared at the wreck.

"You've investigated the truck driver?" Johanna said.

"Yes. Excellent driving record, a solid citizen. Other witnesses saw the accident as well. The truck driver couldn't help himself, really. So, there you have got it. Not very pretty, but then again, this lot's full of similar stories." He gestured toward the rows of wrecked cars.

Goodman pulled up his collar. The cold was starting to bother him. "Why do you think the car swerved, Inspector?"

"I think he suddenly lost control of the vehicle. Why, I don't know."

"Does that suggest the car was tampered with?"

"Not really. To be truthful, Dr. Goodman, if Mrs. Werner had not brought the subject up, we never would have looked for anything like that."

"But you did?"

"As best we could. Look, I'll show you." He called to one of the mechanics. "Horst, lift up this car, the brown Volkswagen." With a small tow truck the mechanic towed the wreck to the grease pit, then raised it on the hydraulic lift so that the underside could be inspected. The inspector pointed out the problems he had in making sure nothing was tampered with. "We checked the brakes, the wheel, steering mechanism. Difficult to do because of the beating it took, but I think we did a decent job. Look, see how we carefully separated the parts." He pointed to the steering shaft, where brightly colored streamers had been placed around different components to indicate they had been scrutinized. "So as best we could determine, there was no evidence that anyone had been tampering with the car." He shook his head. "No, we found nothing." He beckoned for the mechanic to return the car to its place in the lot. Walking back to the station, he again apologized. "I'm sorry we could not be more helpful, but I really think there was nothing wrong with the car."

Goodman stopped. "Perhaps there was something wrong with the driver?"

The inspector shook his head. "Before Doctor Werner died, a full toxic screen was performed."

"Blood and urine samples?"

"Yes, both. The full range of sedatives, barbiturates, tranquilizers, alcohol—I can give you the list inside if you want. It's a standard one. Nothing was found."

"And there was an autopsy?"

"Yes, last night. Mrs. Werner insisted on it."

"And?"

"Nothing," the inspector said matter-of-factly. "Multiple injuries, of course. No evidence of a heart attack or stroke. Again, as best as could be determined under the circumstances. The most likely thing is that he simply lost control of the car for a second and the truck was right there. Probably deeply immersed in some mental calculation or something. He forgot where he was for a moment, made a slight mistake. Any other time, he would have gotten away with it."

"No, no," Johanna insisted. "Peter was too cautious and careful a driver. He didn't do things like that."

The inspector shrugged. "No one is perfect."

109

Goodman watched the mechanic towing the wreck back to its spot in line. It was really not that implausible, the inspector's scenario. Werner was a careful driver—and yet, how many times had Goodman himself forgotten where he was, lulled by the radio, the warmth of the heater, or thoughts of work. Hadn't every driver at some time made a similar error? It was just too bad that truck was there or else he could have regained control and pulled back in his own lane. Or at least if he had bumped into another car, he probably wouldn't have been killed. He turned back to the policeman. "I take it, then, that you're satisfied that there was no foul play, Inspector."

"Quite frankly, yes."

As if mesmerized by the sight of Werner's car, Johanna's eyes had remained fixed on it as it was lowered to the ground with a horrible clanging sound. "What about a gun, Inspector?" she said, still partially facing the wreck. "Was a gun found in the car?"

The policeman nodded. "Yes. We did find a gun in the car. An old pistol, a Luger, probably a remnant of the war. It was like a souvenir, you know, it could not be fired."

"Maybe it was to frighten off someone who was endangering him," Johanna replied. "Maybe that was all he could think of. He was not a violent person."

The policeman shrugged. "What can I say? Maybe so. But for now, it is an automobile accident, no more, no less."

Goodman realized there was no more that they could accomplish there. Taking Johanna by the arm, he gently guided her toward the exit. "Thank you, Inspector. You've been most helpful. We appreciate your taking the time."

"If you should think of anything that could shed further light on this tragedy, please call me." He gave her his card. "I will be glad to listen. I'm a good listener." He smiled, shook Goodman's hand, and walked quickly back to the station house.

In the street near the red Porsche, Goodman stopped. "Well, what do you make of it?" Johanna smiled weakly, "I know what Trudi means. I just don't like the sound of it. You know what a careful driver Peter was."

Goodman had to admit that he had known few drivers as careful as the dead man.

"Maybe there was something wrong with the car that they couldn't find," Johanna went on. "Even the inspector admitted it was such a wreck anything was possible."

"But there didn't *appear* to be any tampering."

"What if something was done to Peter to cause him to have an accident?"

"They tested for poisons and so forth. You heard the inspector."

"Yes," she said wearily. "But what bothers me is that the witnesses say he suddenly swerved. Doesn't that suggest that something went wrong either with the car or with him?"

Grudgingly, Goodman admitted that had bothered him also.

"And the gun," Johanna added. "Why was he carrying the gun? It couldn't shoot."

"I don't know."

"What do we do next?"

"It all depends on how much it will take to reassure Trudi that it was an accident. The police certainly seem convinced."

"But the gun, why the gun? Was it something to do with his job, like Trudi said? How can we find out?"

"On Monday, talk to Richter and perhaps Hoffman."

"Would you do that with me?"

The pleading look in her eyes left him no choice.

London, January 10

"Hello, Dan. Are you enjoying London?" The connection was excellent.

"Shlomo, I'll never complain about the January weather in Israel again."

Shlomo laughed. Good, Tamir decided. The better his mood, the more likely he was to accept his plan.

So far, he had already confirmed his suspicions as to the chemical structure of the nerve gas the Libyans were working on. Now Lester was checking out confidential British military chemical warfare sources for more information and a possible antidote. But even an antidote might not be practical if not administered in time. No, what he had to do, Tamir was convinced, was to see what he could learn from Elizabeth, Dr. Hawthorne's secretary. It had to be more than coincidence that she was the Libyan's English girlfriend and worked in the same institution with Prof. Lester. There had to be a connection. This is what he explained to "Yossi," the young Embassy attache with an impeccable English accent who finally made contact with him late on Saturday, after the Sabbath was over.

The Israeli was skeptical. "We have to run it past Shlomo first. He told me you were to do nothing unauthorized."

"How soon can you reach him?"

"Tomorrow, on the eleventh, about noon Israeli time. On Sundays, he's usually in his office by then. That's his routine for starting the week. Even if he's in the field, they can find him. Meet me at the embassy at 11 A.M. tomorrow."

On a quiet Sunday morning with a fine drizzle falling, the Israeli Embassy on Palace Green—right beside the Kensington Palace and Gardens—appeared deserted. Aside from the guard shack at the entrance to the street, external security was minimal, though Tamir sensed there was more than met the eye. Under the watchful eye of a swiveling TV camera, he pressed the door buzzer, identified himself, and was shown inside. In the telecommunications room, contact was eventually made with Shlomo. Even though he was not in his office, but at an airbase in the Negev, the voice communication was crystal clear. Conversations went back and forth through a scrambler.

"I met with Professor Lester," Tamir continued, "and he confirmed my impression about the Libyan compound. There's a possibility that the British chemical warfare people will have more information. He's checking into it for me, discreetly, of course. In my mind, there's little doubt that it's highly lethal and probably can be spread readily in an aerosol mist."

"Not too encouraging news. But not totally unexpected. I think it's time to move on to Germany."

"There's something else here, though, something we've got to follow through on."

"What?"

"One of the secretaries at Lester's medical school is the girlfriend of your dead Libyan. I recognized her from the picture."

Shlomo let out a long sigh. "You're right, that *is* important."

"I want to question her. The best way is to involve Lester in it as well. She knows and trusts him."

There was silence. Shlomo was thinking. "Alright," he said finally. "But for his sake as well as ours, tell him as little as necessary. When will you do it?"

"Tomorrow at Lester's office. I'll ring him up tonight and explain it as best I can. Remember, Shlomo, he's a friend of Israel. He can be trusted."

Shlomo grunted. "We trust as few of the British as we have to. I'll fax the embassy a copy of the photo. Yossi will see that you get it. You'll probably need it when you confront the secretary. And, above all, don't give her the impression that she's going to be arrested, etc. Most likely, she's a harmless dupe. It's typical of the way the Arabs operate. Good luck." The conversation was over.

• • •

Shlomo sat by the telephone, smoking a cigarette, still thinking about what Tamir had said.

"Good news, Shlomo?" his host, an air force general, asked. "Good enough to call off our little meeting?"

"Not that good, but promising. Not good enough to call off our meeting. But first, I want to see this demonstration you have planned for me."

"Of course, just follow me."

They stepped out of the base administration building into a full blast of hot desert air. Like other airbases in the Negev, this one looked inconspicuous both from the air and the ground. Airplanes and hangars were either under the desert sands or camouflaged to blend in with the scenery. The general spoke into a walkie-talkie and within a few minutes, the sand a mile away from them began to shake. Shimmering columns of hot air rose five feet, ten feet, twenty feet, one hundred feet. With a deafening roar, four F-16s with the distinctive blue Star of David on their wings and fuselage taxied out from their earthen shelters and raced at full blast down the runway. They were airborne in less than 15 seconds, silver wings flashing in the bright sunlight.

"The mockup of the Libyan chemical factory is less than five kilometers from here," the general said. "Take these binoculars and follow them south."

115

"Our new 'smart bombs' are even better than the ones we used to destroy the Iraqi nuclear reactor in 1981. The Americans are so envious that they placed an order for two hundred of them and used them with great success in the Gulf War. But they're damned expensive, I can tell you that." The warplanes flew to the south, circled the target once, then dove sharply and fired their air-to-ground missiles. Explosions racked the desert air. Shlomo watched smoke rise from the debris.

"The flight commander reports perfect hits," the general said with unabashed pride. "This is the way to do the job, Shlomo, a clean surgical strike. No casualties that way."

Shlomo smiled. "Not if you don't take out the SAM sites."

"We'll come in low to avoid them, but still . . . I'd like to take them out, too. I think we've got the Prime Minister leaning our way. He'll approve it."

"I don't think he's made up his mind yet," Shlomo said. "For one thing, he's got the Americans on his back. The CIA has a pretty good idea what you're planning and they're dead set against it. They think Khaddafi isn't using the Rabta plant anymore, so what's the point in bombing it? They're preaching restraint, at least until we find out for sure if the new stuff is being made there. Anyway, that's the purpose of these meetings, to discuss our options. What time will your people be here?"

The general looked at his watch. "They should be here now."

"Then let's get started."

Cigarette in hand, Shlomo followed the general back into the administration building. The four F-16s had already streaked off to practice their low-level attack approaches for the umpteenth time.

Shlomo did not like this kind of meeting. He would be outnumbered by the military five to one. The air force would argue for two strikes, one to take out the SAM sites and one for the plant itself; the army would want to land airborne troops and blow the plant up that way. And what could he tell them? That a raid wouldn't be worth the casualties unless he could guarantee them Rabta had what they wanted? And he couldn't tell them that because he wasn't yet sure. Tamir would be helpful in London, so would the others already in Germany, but there were still too many unanswered questions before he could say with confidence that a raid on the plant would stop the poison gas production. For all he knew, the main plant could be in Iraq—that's what the Russians were intimating. They seemed genuinely worried, which was not the usual state of affairs. In fact, the Russians seemed more scared than the Israelis. Not a good sign. That's probably why the Prime Minister was panicking.

Frankfurt, January 12

When Johanna arrived at Frankfurt Pharmaceutica with Goodman early on Monday morning, she decided she would rather not go in to see Richter. She had spent most of Sunday consoling her sister and was afraid she would not be able to control her feelings with Richter. "After what Trudi told me about the way he annoyed Peter, I'm afraid I might lose my temper and say a lot of things that I'd regret later. Perhaps it's better if you see him alone. I'll walk over to the lab and talk to Dieter Hoffman. Is there anything in particular that I should ask him?"

"This business about the pistol is curious, so see what he knows. You know, Johanna, the more I get involved in this 'investigation,' the more I think you may be right about it being something important. I think I should at least give the appearance of believing that it wasn't an accident. Am I confusing you, or do you understand what I'm trying to say?"

"You mean you might learn more if people thought some sort of foul play had occurred?"

"Exactly. I really don't think there's much to learn, but I think people will be able to vent their feelings more easily if I seem really concerned. That way, when I tell Trudi it was an accident, I'll feel I've honestly pursued it. At first I thought it would be enough to go through the motions, but in fairness to her, I'm going to do it as if I really believed there was foul play. Now, having said that, let's see what Richter and Hoffman have to contribute."

Richter's secretary rang for her boss while Goodman waited in the anteroom. It was obvious from Richter's expression that he was surprised—very surprised—to see him. "But my dear fellow, why didn't

118

you tell us you were staying on? I would have arranged a hotel room, a car." He ushered him into his office, a well-appointed room with beige and white walls and a large corner window that overlooked most of the company's grounds.

"I'm here at the request of Peter Werner's family, Dr. Richter. They're greatly concerned about the circumstances of Peter's death."

"But we all are. My God, he was like a son to me." He sat in a huge swivel chair behind his desk, his heavy jowled face a picture of consternation. "We are all still in a state of shock. The laboratory people constantly talk of how he will be missed. But it was an accident, what more can be done? Is there a shortage of funds for the children?"

"No, nothing like that, not yet anyway. The family wants to know if you're satisfied with the official version of the accident?"

"Yes, why not? Is there anything to suggest otherwise? Everyone knows those trucks are a menace. We have few speed limits and they drive like madmen. Why do you think it could be otherwise?"

"I'm not sure that I do, but his wife and sister-in-law find it hard to accept."

Richter rubbed the side of his chin with his hand. "Yes, I can understand that. Of course, the unexpected nature of his death, the suddenness. What do I make of it? I don't know, I haven't really thought of it as anything else but an accident."

He seemed suddenly subdued, lost in thought.

"What will happen to the drug trials now?"

"With 368, you mean?"

"Yes."

The older man sighed heavily. "For the time being, nothing should change. Stern will continue as before. Hawthorne has been injured and cannot work

119

for awhile. You and Eriksson are not yet really involved. But, you know, Peter was the real guiding force behind these studies, not I. We shall miss him terribly. I don't know what will happen in the long run." He shook his head sadly. "Yes, we shall miss him."

"Yet it was obvious that you and Peter were not in agreement on how the study should be handled."

"Yes, we had our differences," Richter said defensively. "I am more . . . more cautious in these matters, I suppose it's my age." He forced a smile. "I have the company to look after, too. But I think Peter would have convinced me eventually that he had discovered the Perfect Drug. He was a very persuasive young man."

"You once intimated that there were other companies which would like to have 368."

"Quite true. A number of them, in fact."

"Is it at all possible, Dr. Richter, that Peter's death was in some way related to their interest in the drug?"

Richter's eyebrows shot up. "Are you suggesting Peter was killed so another company could get ahead of us on this project? I can't believe that. It's absurd. I know most of the people in this business. Competitors, yes, but not murderers. It's absurd, really. And anyway, it was an accident. Everyone knows that, you should go talk to the police."

"I have."

"Well," he said, growing more animated, "didn't they tell you it was an accident?"

"Yes, but I was thinking of something, perhaps farfetched, perhaps not."

Goodman remembered the conversation with Grundig and his implication that North Korea or other hard-liners wanted to get their hands on the drug.

"What about the Communists? Is it possible they had a hand in this?"

Richter's face lost its scowl. He thought for a moment. "That's foolish. A drug is not like a military secret, you know."

"Yes, but the propaganda value . . . "

"True," the older man agreed. "Let me think about this for a moment. I must say I never considered such a possibility, but perhaps I should have. But how could they stage such a thing to make it look like an accident?"

"I don't know. Someone may have tampered with the car in a way that was not obvious. They may have only planned to hurt him—not to kill him—so that he no longer could be that guiding force you talked about. Maybe someone was trying to hurt Hawthorne in the same way," Goodman added.

Richter seemed interested. "Hawthorne, too? Perhaps someone does have it in for Peter's work—or for the company, for that matter. Yes, you may be right. Somebody may have just wanted to hurt him. Amazing." He got up from the chair and began to pace the room like an angry bull. "Some of those bastards in the East will do anything to retain power, so it's possible they could think of something like this to bring them prestige. There are also many former security policemen who are now unemployed. Some have intelligence backgrounds and are selling their services to the highest bidder for industrial espionage purposes. Others are entrepreneurs: stealing secrets and then offering them to East Germany's former allies. Now mind you, I'm not saying it wasn't an accident, but if it wasn't, yes, any of those types could have done it." He seemed to be warming to the thought. "In fact, the only other firm that's doing research in this particular area is located in East Berlin. There is also a well-known arrhythmia expert in the same city." He searched his mind for the name. "Ach, I can't remember, but Stern and Eriksson know him. The die-hard Communists are a treacherous lot, so anything is possible. Perhaps you are right. Maybe they are somehow involved in this. Poor Peter, to think his death could have been caused by those vermin.''

"I'm only offering it as a suggestion, I have no proof."

Richter waved his hand toward the ceiling. "Of course, of course, I understand. Perhaps I got carried away myself. Well, do let me know what you turn up and, of course, let me know if I can be of any help. Again, please convey my condolences to Peter's family for this tragedy. They're lucky to have a man such as yourself available to help them. I congratulate you on your willingness to be of assistance."

Although Goodman got the distinct impression that Richter was through talking to him, he wasn't about to leave quite yet. "Then you think it's worthwhile to explore this possibility of industrial espionage, especially from the East?"

Richter shrugged. "Worthwhile? That I can't say, but possible, yes, it's possible."

"One other thing before I leave. Peter carried a pistol to and from work. Do you know any reason why he would feel the need to have a weapon?"

Richter's eyes narrowed into two slits. "A pistol? What on earth would he do that for? The plant's safe and always has been. But I can see what you're driving at," he added quickly, "perhaps if someone had threatened him, one of *their* agents, for example, he would need the weapon as security. I certainly can't think of any other reason for him to have a gun."

Goodman left Richter to puzzle over that last bit of news. He walked back to the Porsche, but Johanna had not yet returned. Despite the cold, he decided to walk around the grounds; the small lake behind the plant especially intrigued him. It was situated

122

in isolated splendor surrounded by a smattering of now barren oak trees. Two wrought iron benches were placed by its edge and a stone footbridge straddled its narrow center. The surface of the pond was frozen, the ice glistening in the bright sunshine. From the bridge, he saw how clean and sharp the lines of the plant were and how the huge glass windows caught the sun's rays and reflected them back in a flash of light. It was indeed an architectural delight. Unfortunately, Goodman had no more than a few minutes to enjoy the view from the tiny bridge before he noticed a security guard racing across the field toward him.

"You've got to get off that bridge at once!" the guard yelled. Seeing Goodman's puzzled look, he explained politely but firmly. "Those are Dr. Richter's orders. No one is to use the bridge, it's not safe. I must ask you again to get off it at once."

Goodman was annoyed. "How was I to know?" Goodman said, as he walked back to the edge of the pond, "there are no signs."

"All employees know the rules. We have so few visitors there seemed little point in putting up ugly signs."

Goodman laughed. He had not expected this rather rough-hewn fellow to be so aesthetically minded. "All right, I've had enough sightseeing, you've made your point."

By the time he reached the red Porsche, Johanna was inside with the motor running and the heater turned to HIGH. She rolled down her window when she saw him. "I had no luck," she offered, "Hoffman was off somewhere in the building and I waited as long as I could before giving up. How about you? Did you see Richter?"

"I did. Let me join you in that warm little car of yours and I'll tell you all about it."

While he made himself comfortable, she pointed to a window on the top floor of the building. "That was Peter's office. Sometimes, he would ride in with Hoffman and Trudi and the kids would come down in the afternoon and pick him up." She imitated their waving to him. Saddened by the scene she had recreated, she sighed and squeezed his arm tightly. "You were very good to go in there and talk to Richter. Thank you. Did you learn anything?"

"Only that Richter is intrigued by the idea of industrial espionage, especially if Communists or ex-Communists are in any way involved. And I think the news of the pistol really jolted him. He was almost speechless, probably for the first time in his life."

"Do you really think the Communists could be involved?"

"I was only floating a trial balloon. I still find it hard to believe that Peter's death was anything more than an accident, but, as I said, if I'm going to play the devil's advocate, I have to be convincing. I even threw in Hawthorne's robbery as part of it."

"How can we find out more about possible Communist involvement?"

"Richter told me that there's an East German firm that might conceivably have an interest in Peter's work. I think I know the people involved, and so does Eriksson. I think I'll call him."

"Do you know how to reach him?"

"I know what hospital he works at. That should be enough."

"Let's do it from Trudi's house." She gunned the engine and sped off.

● ● ●

From his office window Richter watched the Porsche speed off. His face showed no emotion, but inwardly

124

he was annoyed. Goodman was being a nuisance. Why this persistence about Werner's death not being an accident? And this news about a pistol in the car. Why did Peter need a weapon—how much had he really known? Richter shook his head. He had no answers to these questions, nor to an equally disturbing one. Why was Goodman snooping around the pond? Was he simply the research scientist he appeared to be, or did he have other reasons for his Frankfurt visit? Richter frowned. You're getting upset over nothing, he chided himself. On the other hand, his suspicious nature had helped him avoid danger in the past; might it not be warning him again? He sighed wistfully; things had been so quiet for so many years that he'd been lulled into a false sense of security. He had become a much more deliberate person, one who did not like to be rushed, to take chances unnecessarily, but this fellow Goodman was spoiling his careful plans. He could see trouble coming. He would have to be very, very careful.

• • •

Eriksson was surprised to hear Goodman's voice. "Where are you calling from, Paul?"

"Frankfurt."

"What in the world are you still doing there?"

"Peter Werner's dead. I've decided to stay a few more days."

"Peter's dead? I can't believe it. When did this happen?"

"Several days ago. Automobile accident, supposedly."

"Connection's bad. Did you say supposedly?"

"Yes." Goodman related the story of the accident.

"But why are you involved? Aren't the police enough?"

"The family asked me to look into it."

"Well, what do you think? Is there anything suspicious?"

"Probably not, but there is one aspect that deserves some checking. As you recall, Richter is obsessed by the Communists. He knows Grundig's firm is interested in the drug. Of course, I said nothing about our meeting with him, or his desire to get his hands on the drug. How Peter's death would help Grundig is another question."

"Rubbish. Why would they want to kill him?"

"Perhaps just to hurt him, or scare him enough so that he couldn't work for awhile. That's what seems to have happened to Hawthorne."

Eriksson grunted. "Funny you should mention it, because I *have* wondered about that. Apparently the thugs that attacked him had little interest in money. They just wanted to give him a beating. Methodical, too. I wondered if someone was settling a grudge, but I never connected it with 368."

"Well, Hawthorne was very enthusiastic about the drug. More than anyone else. Now presumably he can't work on it—or anything else—for awhile."

"That's true. Actually, the only person who would be pleased about all of this in a perverted way, of course, is Richter, and I would think even he's got more feelings than that. But what do you propose? Another meeting with Grundig?"

"Exactly. If anyone from the East is involved with 368, he would know about it. Not necessarily from the dirty side of things, of course."

"If you like, I'll give him a ring and feel him out very gently. Give me your number and I'll ring you up when I contact him."

"Thanks for your help."

"For Peter Werner, it's the least I could do."

There was someone else Goodman wanted to talk to: Werner's chief technician. He dialed the plant and was connected to the laboratory. Dieter Hoffman was now back at work. Goodman explained his purpose in calling. "Any information that you could give me regarding Dr. Werner's activities in the last weeks might be helpful, especially if there was anything out of the ordinary. Did you notice if he had a pistol?"

Hoffman wanted to be helpful, but could add little. "Dr. Werner followed the same routine as always. I noticed nothing especially different, certainly not a pistol."

"Was he more depressed, or unhappy lately?"

"I would say he had been out of sorts for several months. This was nothing new."

"What did you attribute that to?"

"The new drug, 368."

"Was he worried about its formula being stolen? By East Germans, for example?"

"No, he was just edgy. Nothing specific. He said nothing about East Germans. Of course, his sessions with Dr. Stern didn't help."

"What sessions?"

"Dr. Stern would drop in periodically to check the progress of the 368 study. Once in a while they would have loud discussions."

"Arguments?"

"I'm not sure if I would go that far. They did have one of these discussions the day before the accident. But Stern came back the next day and he was all smiles."

"What did they talk about?"

"About whether Stern's results were misleading because of inadequate dosage. Dr. Werner told him to increase the dose. Stern was afraid of side effects. Then Dr. Werner told him about his own separate studies—the ones he told you about earlier. Oh yes, I know he told you. Anyway, Stern wanted to see those reports, but Dr. Werner said no. That's when they raised their voices. 'How can I believe you when you keep things from me?' Dr. Stern yelled, and he stamped out very angrily. But the next day, he was back and in a better mood. Yes, he even brought a peace offering for our afternoon tea. Delicious sweets, all of us in the lab had some. They seemed to put us in better spirits."

"This was about 4 PM Thursday?"

"Yes."

"And Peter left for home at 6 PM as usual?"

"So I understand; I had gone a little earlier."

"Where are the reports that he wouldn't let Stern see?"

"That I don't know; I got the impression he kept them at home."

Goodman was confused. What did this business with Stern mean? He wondered if there was any significance to his being virtually the last person to see Werner alive. Could it really be that Werner's death was more than an accident? Goodman wanted to clear his mind of any further suspicions of foul play but then remembered how Richter became excited over the Communist plot theory that he had dangled in front of him. And the pistol business. And the lingering doubts about Stern. He wondered if Stern was somehow involved in Werner's death. And what about Meyer?

When he mentioned Meyer's name to Hoffman, the laboratory assistant cursed out loud. "The son-of-a-bitch is gone, you know. Went from his skiing holiday straight to East Germany."

"East Germany? Why would he do that?"

Hoffman paused. "The old Eastern zone has become a haven for neo-Nazis and similar types. I don't really want to talk any more about Meyer. There's a lot about our company and the people that work in it that you don't know, Dr. Goodman—and you're probably better off not knowing about it."

"Don't talk in riddles. Peter Werner's dead, remember? Maybe Meyer had something to do with it. He seemed very touchy when I mentioned 368 to him."

"Meyer did more important things for this company than work on 368. If you want to know more, you can ask Richter. But don't tell him I tipped you off. That man's got a bad temper." Hoffman hung up the telephone. End of conversation.

Meyer and Richter and Stern, Goodman mused, what an odd collection. And then there was Grundig.

Restless, he wandered into the living room, where Johanna, Trudi, and the children had gathered. He made small talk as best he could, but it was awkward sitting in Werner's house talking to his widow while the children played all about them. Memories of their father constantly intruded on the conversation. Trudi was still fragile, eyes tear-filled at the slightest provocation, no make-up on, clothes that didn't match, hair untidy. Now she was doubting her suspicions, wondering if the quest she had started Goodman on was fruitless. Then he recounted his conversation with Richter. At the idea that perhaps the Communists were behind this, her eyes lit up.

"They did things like this. Every month, the newspaper used to have some dreadful account of what they have done. They had spies everywhere. Maybe some of them are still active."

"I thought those were mostly political things," Goodman said.

"Yes, but there *were* also reports of industrial spying and scientific spying."

"Anyway, I have some feelers out via a colleague in Denmark."

"Eriksson?" Johanna asked.

"Yes. Stern's name seems to crop up frequently as well." He recounted Hoffman's story of Stern's visits and the arguments.

Trudi smiled triumphantly. "See, I told you something at work was upsetting him. See, Dr. Goodman, I was right. Now you understand."

"The nub of it is that Peter didn't really think Stern was doing a good job on the research. I think that's why Peter secretly supported another series of studies. But why Stern would be antagonistic, I don't know. Incidentally, I would very much like to see any records that Peter may have kept at home. I have a feeling that he didn't keep all his material at the plant. I'm especially interested in seeing who the doctor was that helped him do his studies. He may know more about why Stern's results weren't better or at least why Peter thought they should have been. Isn't there an area around the house where your husband worked?"

Trudi nodded. "Downstairs in the corner of the children's playroom, he had built a desk and some file cabinets. If you think it's important, we could look through them."

"Would you mind if I helped? I might recognize it more easily than you would. I know there may be personal things as well, so if you would rather I didn't . . . "

"No, no," Trudi said, "you are our friend. Please come with us."

For the next several hours, the three of them opened the file cabinets and went through each folder, looking for any mention of 368. There were several 24-hour electrocardiogram records, but they were clearly marked with a different drug number and were faded. Presumably, one of 368's predecessors. Working

130

with her husband's "at home" files proved too much for Trudi, and she soon excused herself to go back upstairs.

"She seemed to perk up a bit when you mentioned the Communists," Johanna said.

"Are they that hated?"

"The police and government officials? Yes."

After another futile half-hour, Goodman had had enough. "Maybe we can do more tomorrow. I'm exhausted." Johanna agreed and they put everything back where they had found it. Before going upstairs, Goodman looked about the room at the bookshelves lining the walls, the fireplace, the children's toys, and posters on the walls. Where would Werner have hidden something he didn't want to be found, if indeed he had hidden anything?

"After dinner, I'm coming to your room," Johanna said as they walked up the steps. "Unless you're still too tired."

Goodman grinned. "I'm regaining my strength already."

Dinner at Werner's home was no longer a happy time. Trudi's meal had not been well prepared and they had gone through the mechanics of eating. The children were irritable and the mood of the house was oppressive. The children were put to bed soon after dinner and Trudi followed soon after. Johanna and Goodman were alone in the living room.

"I've missed your body next to mine," she said. "I didn't want to tell you that to inflate your male ego. Any more, that is. But I can't help myself." And with that, they made their way quietly to the guest room where he had slept.

He stood behind her and put his arms around her waist. "I missed you, too. I stayed on because I wanted to help, of course. But I also wanted an excuse

to be with you. I just wish it hadn't been this way."
Burying his mouth in her neck, he inhaled perfume
mixed with the smell of her skin. Her presence excited
him as much as it had the first time they were together.
She tensed slightly, then relaxed and took his hands
and placed them over her breasts. "I missed you so
much," she said. He flicked off the lights and they
undressed quickly.

"I wish I could lie here all night," she said after
they had made love, "but Trudi has a habit of coming
into my room looking for solace. I don't want to upset
her by not being there."

He stroked her back and thigh. "Does Trudi
talk about us?"

"Not much."

"She doesn't mind then?"

"If she does, she isn't saying."

Moving closer to him again, she rested her
head on his chest. "Those few days away together were
heavenly. I shall always cherish them."

They stayed in bed, talking until she decided it
was time to go back to her room. He watched her dress,
then kissed her good night. As soon as she left the
room, he fell asleep. It had been a long day.

During breakfast, Eriksson called back. "I'm
sorry to call so early, but I did want to catch you
before you went out. I have some interesting news
for you."

"Interesting or good?"

"Both. We had a stroke of luck. I rang up Professor Grundig in Berlin, but he wasn't there. He was
on a lecture tour in France. His secretary told me his
plans call for him to come back through Frankfurt today. He has a talk at the medical school at noon, then he
leaves for Berlin tomorrow. He'll be staying at the
Grand Hotel. That gives you a little time to try and

132

contact him, get a chance to talk to him again and see what he really knows."

"Fantastic. I hope it wasn't too much of a bother for you to find all this out."

"Nonsense, I was glad to help. Just a word of advice. Grundig has a good thing going and he wants to keep it that way, especially if the North Koreans are financing him. Based on our last meeting, I don't know if you can believe everything he tells you."

"I figured that one out myself, Lars. Thanks again. I'll try to keep you posted on what happens."

"Right. Bye. And be careful. First Hawthorne, then Werner, it's not an encouraging trend."

After breakfast Goodman decided to continue to look for material Werner might have left around the house. He tried to put himself in Peter's shoes to picture where he might have hidden important data if he had lived in that house. The children's playroom had toys, more toys, a desk, fireplace, bookshelf. Books. Why not fold the reports and put them in one of the little-used books? He and Johanna went back to the playroom. Methodically, they emptied all of the books on the shelves. When they were nearly through with the last section, Johanna called out excitedly, "I've found something!" She brought over two sheets of paper. They were summaries of 24-hour electrocardiograms on two different patients. Each sheet was divided into two parts: on the top were the control or baseline findings; on the bottom were the same patient's results after 368 had been given (the number of the drug was clearly marked in red and circled). Both patients showed a complete absence of irregular heart activity after they received the drug.

"This is what Peter was so excited about," Goodman said. "This shows the drug works and works well . . . "

133

"They're better than the results in Stern's patients?"

"Much better. It's only two patients, but I see now why Peter was so optimistic. These clinical studies confirm the animal experiments. I wonder who these people are and if they're taking the drug now. But more importantly, who is their private doctor?" He pored over the papers. "There's nothing on these sheets to give any clues."

"Shouldn't their names be entered somewhere in the records?"

"They should be, but they're not."

"Maybe there's another set of reports down here." She quickly went through the remaining books, but no more papers fell out. Frustrated, they began looking behind pictures, even in the children's toy box. Nothing.

With Trudi's permission, they went through the rest of the house, but still nothing turned up.

"The doctor must be in Frankfurt," Goodman reasoned. "Who were Peter's close friends in the medical community?"

"There were dozens," Trudi replied. "I wouldn't know where to begin."

"Those who come to the funeral tomorrow are a likely group to start with. I'll see who shows any interest at the mention of 368. I'll also check with Hoffman."

Hoffman had no suggestions. But Goodman was in luck when he checked the Grand Hotel. Grundig was in and agreed to see him at the hotel in an hour.

"Do you think you can trust him?" Johanna asked.

"What choice do I have?"

• • •

Goodman had to make his way downtown without a car. Johanna had noticed the Porsche had developed a problem in slowing down at high speeds and suspected a malfunction in the brakes. While she had it looked into at the local garage, they were temporarily carless.

"You shouldn't have any problem," she explained, "we have an excellent express bus system. Walk a few blocks to the square with the post office and you'll see large signs indicating the bus stop. It goes directly into the central part of Frankfurt near the old city hall. From there you can walk to Grundig's hotel."

Taking a bus in Germany was a new experience for Goodman, but he followed Johanna's direction and had no difficulty in finding the right one. Once inside the bus he tried to concentrate on the views outside the window but couldn't because he had the uneasy sensation that someone was watching him. Whenever he looked at the other passengers, their eyes were always elsewhere, yet the uneasy feeling persisted. When the bus finally deposited him in the downtown area and he became a pedestrian again, the suspicion lingered. A quick stop and a turn to look back were to no avail: no one seemed the least interested in him. Why am I so jumpy, he thought, wondering if this investigation he was pursuing was making him paranoid. Disgusted at himself, he walked straight ahead, not stopping again until he reached his destination, the Grand Hotel.

The Grand Hotel was one of Frankfurt's newest and most elaborate structures. It stood in stark contrast to the Hotel Fürstenwald and many of the other older and smaller buildings in the neighborhood of the town hall. Inside, it was as ornate as one would have expected from its garish exterior. To find one of the former German Communist regime's most celebrated scientists ensconced in such luxury was a bit ironic to Goodman, but to Professor Grundig it was not at all out of place.

"I've always traveled first class, even in the 'old' days. The state appreciated my services," he explained in the sitting room of his suite. "They knew I had to be comfortable when I traveled. There is really no inconsistency in that. After all," he smiled broadly, "I came back every time then, and I still do now." He took off the thick eyeglasses with black frames and stared at Goodman quizzically. "Have you had second thoughts about my offer?"

Goodman shook his head. "No, I can't help you out. I thought you had other ways to get what you wanted?"

"Unsuccessful, unfortunately. At least, so far. But if you aren't going to make the drug available, why this meeting?"

"I want you to tell me what you know about Peter Werner's death." Goodman said steadily while he returned the East German's stare. "I presume you've heard about it."

Grundig's tone softened. "Yes, of course. Very sad news."

Goodman leaned forward. "What can you tell me about it?"

"I don't know anything and even if I did, you're someone that I hardly know, and whose motives are unclear. What exactly are your motives, Dr. Goodman? Do you want to work on this drug in your own laboratory now that Werner is dead, or is Richter going to hire you to work for him?"

"No, it's nothing like that."

"Good. He's not a very trustworthy person."

"How so?"

"In many ways."

"Why can't you be more specific?"

Grundig leaned back and studied the younger man carefully. "For reasons that I don't care to discuss at the present time, perhaps later. You'll just have to take my word for it. That is all I can say about the matter now."

Goodman didn't know whether to believe him or not.

"Professor Grundig," he persisted, "I'm afraid I must be direct: Do you have any knowledge of any attempts to harm Peter Werner?"

Grundig showed no surprise at his question. He drummed his fingers on his knee for several minutes before speaking.

"Scientific espionage is not unheard of in this part of Europe," Grundig finally answered in a sober tone, "and there are many of my ex-'countrymen' involved in it, but in all honesty I know of no such activities connected with Peter Werner's death. You can believe me or not. Even if you do, does that exclude the possibility? Of course not, but the source would not be one sanctioned by the East. I take it that you are very concerned about the circumstances of his death?"

"Yes."

"Perhaps I can be of some help. Let's take a 'little walk.' " He withdrew an overcoat from the closet, and beckoned for Goodman to follow him.

They left the hotel and walked a few blocks to the reconstructed sentry house of the 1700's, the Hauptwache. "They have an outdoor cafe here in the warmer months," Grundig explained. "When I am in Frankfurt, I often stop by for coffee or beer, depending on my mood at the time. Today, my mood is pensive because of our discussion and I would choose the coffee if the cafe were open." He resumed his steady pace, Goodman at his side.

"Do you know where we're going?" Grundig asked.

"We're headed toward the river."

"Correct. But we won't reach it. There's some other place I have in mind." He negotiated his way through the city's medieval center, all largely reconstructed since the war, not minding the cold air, but seeming to thrive on it. Goodman estimated his age at about sixty and marvelled at his stamina. Methodically, the older man led him through small streets until finally he found what he was looking for. "Aha, just as I remembered it. You look startled, Dr. Goodman. Do you know this place?"

Goodman nodded. They were standing in front of the Nightingale nightclub. "I was here last week with Peter Werner."

"I see. Well, one of the employees is an important person for you to talk to. He used to work for Frankfurt Pharmaceutica, but was forced to leave because of his socialist views. Peter Werner helped him get this job. If there is any possibility of foul play in his death, this man would either know about it or know how to find out about it."

"A fellow with a missing arm?"

"Exactly. Then you've already met Heinz Becker?"

"He's the headwaiter."

"Is he? Wonderful. Then this new job turned out very well for him after all."

Grundig went to the door, but it was locked. "Too early for a nightclub to be open, I'm afraid. Come back tonight and talk to him. Maybe he can help you." He held out his hand. "I think this is where we part, Dr. Goodman. It was a pleasure meeting you, and please be assured that if Peter Werner's death was not an accident, my colleagues had nothing to do with it.

However, we are still interested in the new drug, 368. Its importance cannot be overestimated. I wish I could say more, but . . . If your feelings should change, don't hesitate to call me directly. Here is my card. Oh, by the way, just to set your mind at ease, I also had nothing to do with the recent arrival in the East of Hans Meyer. Good-bye, Dr. Goodman."

"Wait a minute, what more is there about 368? And how did you know about Meyer?"

With an enigmatic twinkle in his eye, Grundig strode off in the direction of his hotel. Goodman tried the locked door one more time, then hailed a taxi to take him to the bus station. He had done enough walking for one day. As the taxi passed a news kiosk on the corner of the street, a bearded man in a leather cap and loden coat caught his eye. He thought it was the Romanian "Anton Ionescu" but couldn't be sure. By the time he looked back the man was gone and he wondered if his imagination was working overtime.

• • •

Goodman found Johanna in a state of excitement when he returned from his visit to Grundig.

"We've found the doctor that Peter used for his studies. He called to offer condolences and on a hunch, I mentioned 368. He'll see us whenever we get to his office, which isn't far from here."

"What about your car?"

"All taken care of."

"That was quick service."

"I've learned to use my female charms to get things done just as you taught me in Herzberg. Anyway, it wasn't a major repair job, just a leak in the brakes so that most of the fluid was gone."

"You must have hit a rock somewhere."

139

"I suppose so, though the mechanic commented that the perforation seemed very regular. He thought that was peculiar."

"Peculiar?" Goodman's face tightened.

"What's the matter?"

"The mechanic finds the hole in the fluid line 'peculiar.' It occurs to me that someone may have tampered with the Porsche. We could have been in danger if we were going at high speeds and the brakes failed."

Johanna laughed. "You're making a mountain out of a molehill. It was most likely a sharp rock just as you said."

"I hope so," Goodman said, "I'm not really eager to contemplate alternative explanations. Let's go see the good doctor."

It was approaching twilight as the red Porsche roared through the streets of Frankfurt once more, this time into a fashionable residential area where professional offices were tucked into small enclaves of well-kept homes. Werner's friend, Jergen Friberg, was a small thin man, young but very studious in appearance. Compared to Richter's office, his was compact and efficient without frills. Goodman quickly explained the situation to him, mentioning only briefly the possible involvement of the Communists. Friberg listened attentively.

"I am very glad you're looking into Peter's death. To tell you the truth, I was very disturbed about it myself. I don't know whether or not someone is after the drug, but I do know that Peter has been very much on edge for some time after certain activities at his plant. Although what I did for him was very irregular, we are old friends. He needed immediate confirmation that his animal studies were correct."

"And that you provided?"

"Yes, with two of my patients who had not yet been treated with any other drug."

140

"Did he say anything as to why he didn't trust Stern?"

"He gave the impression that he thought Stern was not a good investigator."

"Nothing more?"

"Nothing definite." Friberg frowned. "I can't put my finger on it. It was something he didn't want to be pressed about."

Goodman turned to Johanna. "I wonder why he picked Stern in the first place. There is a fine research unit in Hannover, if he wanted to use someone from that region, but Stern is not associated with it."

Friberg interrrupted. "Oh, there's no question about that. Peter had told me Richter picked Stern. Richter and Stern had many friends in common, especially Otto Strumpf, Stern's uncle. Peter thought Richter was helping Stern's career by getting him involved in a good research project."

"Otto Strumpf," Johanna said. "That name's familiar."

"He made the headlines several years ago. Someone accused him of using his factory during the war to profit from slave labor, from the concentration camps."

"That's right," she remembered, "but nothing came of it."

"No, apparently not."

"There are a lot of industrialists like Strumpf still around and making a good living," Johanna said bitterly. "They were as bad as the Nazis."

Goodman wasn't listening. He was wondering if it was time to have another talk with Richter. There were too many issues he was puzzled about. He still didn't understand why Werner choose to be secretive about his actions or why he had questioned Stern's reliability. There was also the strange encounter with Meyer—Richter's "boy," and why he soon after had fled to the East.

Friberg had been as helpful as he could, but there was nothing more to learn from him. Johanna drove Goodman back to the plant and waited while he went to see Richter. This time, the secretary's announcement of his arrival brought no prompt response. It was almost the end of the workday and Richter kept him cooling his heels for nearly twenty minutes. When he finally was ready to receive him, he had his secretary bring him in while he remained behind his huge desk. "So, Dr. Goodman, two visits in one day. What have you learned since we last spoke."

"I've had a busy day. I met with Professor Grundig from Berlin."

Richter's eyebrows raised slightly. "Good for you. What did he have to say? Denied any involvement, I'm sure."

"Yes."

"Well, what did you expect? A confession? Now, mind you, I'm still not convinced Peter's death was anything more than an accident. What else did he say?"

There was something about the way Richter looked at him—the intense, inquisitive glint of those steely blue eyes in the sea of red cheeks and forehead—that dissuaded Goodman from saying anything, at least for the moment. Certainly he had no intention of mentioning the references to the headwaiter at the Nightingale. "He said nothing else of importance," Goodman replied matter-of-factly.

Richter relaxed. "Well, so no surprises after all. Perhaps now it's time to forget this whole business. Not that they couldn't have been involved, but how would we prove it anyway? I suppose you'll be going back to New York now?"

"There are still some things I'm not clear about. Perhaps you could shed some light on them."

"If I can, of course."

"It's not at all clear to me how Dr. Stern came to be involved in the 368 studies. There are certainly better qualified physicians available."

Richter's face grew redder. "There was nothing wrong with using Walter Stern. It's true he was my choice and not Peter's, but why not? After all, I am the president of the company. I can pick and choose whomever I want. Peter just didn't like the results Stern got."

"There's been some talk that it was because of mutual friends rather than Stern's ability that he got the job."

"Who said that?"

"What does it matter?"

"What friends did they mention?"

As soon as Goodman said the name, he regretted it. "Otto Strumpf, for one."

Richter's mouth popped open. He was momentarily speechless.

"Otto Strumpf is a dear friend," he finally stammered. "I think this discussion has gone far enough when my good friends are slandered."

"I didn't slander . . . "

"Enough, Dr. Goodman. Enough of this conversation and as far as I'm concerned, enough of this investigation. Good-bye." With that, he wheeled around in his swivel chair and busied himself with some papers on the shelf behind him. Goodman let himself out of the office.

● ● ●

When the door shut, Richter pushed the intercom button on his desk phone and said to his secretary, "Get me Dr. Stern in Hannover. If his office says he's out, have them find him. It's important." He slammed the receiver back on its cradle. Damn that Goodman! What business does he have prying into Otto Strumpf's affairs? He'll soon find himself in the middle of everything. All of which could be troublesome, Richter concluded. Very troublesome. He leaned back in his chair and rubbed his eyes. Was he getting too old for these games, he wondered. Perhaps it was time to consider retirement, perhaps go abroad. He and the company had done its duty long enough. Now let some of the others serve the Fatherland, the new ones, the young ones. Enough is enough.

● ● ●

Outside the plant, the afternoon sun had long since set. "You were in there quite a while," Johanna said a bit petulantly.

"He kept me waiting," Goodman explained apologetically. "And then he lashed out at me for casting aspersions on Walter Stern and Otto Strumpf. I never even got around to Hans Meyer. As far as he's concerned, I should pack up and go home."

"Is that what you're going to do?"

"No. My curiosity about this business is aroused. I think we'll go back to the Nightingale and have a talk with the headwaiter."

"Now?"

He looked at his watch. "They should be there getting ready for the evening show. We might as well do it now."

Johanna gunned the motor of the Porsche, made a quick U-turn, and headed downtown.

The interior of the Nightingale was dark. Busboys arranged tables while the musicians who were to provide the evening's entertainment rehearsed. To Goodman's disappointment, this was not one of the

144

1940s American swing nights; instead, the small group played a succession of German popular songs. Johanna and Goodman peered around the room looking for the headwaiter. Finally, one of the busboys approached them.

"We're looking for the headwaiter, Herr Becker," Goodman explained.

"I don't know if he's here, but I'll check." The man's accent was one that Goodman hadn't heard before. "Where's he from?" he asked Johanna, while the busboy scurried off.

"He's an Arab. Our country is full of them, nearly all doing menial work."

The busboy returned with a waiter.

"I'm sorry," the waiter said, "but Herr Becker is not here this week."

"Oh? Is he ill?"

"No, nothing like that. He's just taken some time off. Is it something important?"

"Actually, it is. Is there any way to reach him?"

"He's gone to his farm in the north. It's fairly well isolated and no phone. But he'll be back next week."

Johanna spoke up. "This is a rather important matter involving a good friend of his. If you told us where the farm is, perhaps we could see Herr Becker there."

The waiter was skeptical. "What do you want with him?"

"I told you. It concerns a friend of his who died recently, Peter Werner."

"Peter Werner?" the waiter said in awe. "Of course, why didn't you say so? We all knew of him." He wrote directions on a piece of paper and handed it to Goodman. "It's near Bergen." Excusing himself, he returned to his duties.

Outside, Johanna and Goodman discussed what they should do.

"Bergen is about a four-hour drive from here," Johanna said. "Maybe we should wait until he returns from his vacation."

"No, that's too long to wait. Is Bergen near Hannover?"

"About an hour further north."

"Why not kill two birds with one stone? We'll see both Stern and Becker."

Johanna liked the idea. "Tonight or tomorrow?"

"Tomorrow's soon enough. Right after the funeral."

London, January 12

Lester's message had been brief and to the point: "Have set up appointment with Elizabeth for Monday. Suggest we meet first in your hotel restaurant for 8:30 breakfast." With these instructions in mind, Tamir left his room a few minutes before 8:30 and was about to walk down the one flight of stairs to the lobby when he heard the chambermaid cursing loudly. He peered into her work closet and seeing she was alone asked solicitously, "Are you alright?"

The chambermaid, a white-haired lady with a red face, shook her head yes. "Oh, I'm fine, sir. It's just that these bloody Arabs are too much for me." Her Irish brogue made the word "bloody" seem twice as long. "The things they do is disgusting. It's worse than the IRA; I don't care what anyone says."

"Has something happened?"

"Didn't you hear about it on the radio, sir? I heard it on the BBC just a few minutes ago. Some Arabs shot up one of the international arrival buildings at Heathrow. Killed two people and injured ten others."

"Did the radio say which group was responsible?"

" 'Anti-PLO factions' is what the radio said, sir, that's all. They're thought to be supported by Libya, the announcement said, or some such words. But all these 'factions' are the same, if you ask me. Bloody lot should be kicked out of the country."

Not a bad idea, thought Tamir as he left her, still cursing.

Downstairs, Tamir found Lester waiting for him in the lobby and they went in to breakfast together. Lester had already heard about the airport incident.

147

"There are a lot of crazies out there," Lester said over his glass of orange juice. "They'll do anything to prevent any possible reconciliation between the PLO, moderate Arab states and your country."

"Someone has to bankroll them," Tamir said, "and protect them. And if a government does it, then that government is vulnerable for reprisals."

"You're talking about Libya, Iran, Iraq or Syria. One or another seems to be behind most of the terrorist groups in the Middle East."

"That's exactly who I meant, but I wonder why they chose Heathrow, and why now? Perhaps it's a warning to you British to stay neutral or more will follow."

Lester put down his glass. "That's entirely possible. It's one of the things we should talk about. And, of course, this business with Hawthorne's secretary. As soon as the waiter leaves."

As if on cue, the waiter arrived, took their orders, and hurried back to the kitchen. Lester continued. "I had a nice chat with some of my friends in the military—chemical warfare people, that sort of thing. They believe the formula that you showed me would be the main part of a very toxic compound. In other words, it is just as you suspected."

"I don't know whether to feel pleased or not, considering the consequences."

"I'm afraid there's more," Lester went on, "and I guarantee you won't be pleased in any sense of the word."

Now it was Tamir's turn to put down his half empty glass of juice. "I'm listening."

"The formula that you were interested in is apparently a minor variant of the standard nerve toxin that is used in most 'poison-gas' concoctions. This was the main chemical weapon produced at Rabta by the Libyans with the help of a West German chemical firm,

148

which has since been fined and its president imprisoned for violating German export laws. The Iraqis also have quantities of it and probably used it on the Iranians during their war. Although there is no real antidote, precautions can be taken and, most importantly, its use is almost instantly detected. My chaps don't think they would use it against Israel unless they were prepared for all-out war. Iraq may be so inclined, but I doubt the Americans would let them try anything like that. On the other hand, the Libyans are out of the world spotlight for the moment and may be jealous. Their actions are hard to predict, but NATO intelligence has let Tripoli know that it is aware of the supplies the Libyans have received. Any use by them would be viewed as equivalent to making Libya a true outlaw nation."

"So far I haven't heard anything that bad," Tamir said, puzzled. "I take it then that there's more?"

Lester leaned across the table and lowered his voice. "It's the *other* formula on that scrap of paper that my chaps are concerned about."

"The other formula? For the cell contraction stimulant?"

"My dear boy. That 'stimulant' is in reality highly toxic to the heart. In fact, it will stop a human heart within 30 seconds of injection. Our chemical warfare people feel it would be relatively easy to convert into an aerosol and as such, it would be devastating. It could be introduced into water supplies but, unlike biological weapons, wouldn't cause an obvious disease. Without knowing how to test for it in the body, we could never prove that it was used as a poison. While you're digesting that piece of bad news, there's one other interesting bit of news which you—and your *military* friends—should know about if you're planning to interfere with the manufacturer of this compound."

149

"What's that?"

"Not only is the compound itself highly poisonous, but so are its breakdown products. So to destroy the compound would be very difficult, since anyone coming into contact with any of these substances would probably die instantly, before the substances had a chance to either be diluted by the atmosphere or to become inactive. Again, to be blunt, merely blowing up the plant that makes these products would be extremely dangerous to anyone in the factory or any civilian population nearby. It would make the incident in Bhopal, India seem trivial in comparison. I think the only way to neutralize this compound is to stop its synthesis: prevent the plant from obtaining the raw materials needed to produce, or physically remove the materials if they've already been delivered to the plant. You really need detailed plans of the plant if you're going to go after it. The other option is to find out who's supplying the raw materials and go after them. It would take a pretty sophisticated outfit, most likely European. Believe it or not, it doesn't necessarily have to be an Eastern bloc country. Lots of Western businessmen would love to get their hands on that Libyan money. Or they would just want to see Israel destroyed period, with or without the profit motive."

On that somber note, Lester began eating the bacon and eggs that had just arrived. Tamir picked at his buttered toast. He was still focusing on the Englishman's earlier remarks. What Lester was saying meant there were *two* possible poisonous compounds the Libyans were producing with German help. Because the heart poison was so toxic, there would have to be a special team in that plant to dismantle the chemical weapons areas. Shlomo would need to know that.

"Now, after all that grim news," Lester continued "let's talk about something more cheerful. What's all this about Hawthorne's secretary and some Libyan army officer?"

"The subjects may well be related, unfortunately. It's possible the Libyan cultivated the secretary to learn what he could about British chemical warfare plans. In other words, the Libyans probably would like to know the same sort of things that I've been asking about, but for a different reason, perhaps to learn how to protect their plant."

"Yes, but why not go after *my* secretary?" Lester asked. "She knows a lot more than Hawthorne's. But then again, she's happily married, so that wouldn't work."

"It's hard to believe it's mere coincidence that of all the women in London, this Libyan officer is having an affair with Hawthorne's secretary."

"Well, let's go talk to her, then. I take it you want me to do the questioning, to make her feel at ease?"

"Correct. I'll brief you on the ride out to the medical center on what we've found out about her boyfriend. We just have to stop at the Israeli Embassy, which is very close, to pick up a photograph that may help you overcome her defenses."

By the time they reached the East Acton underground station, Lester had been led through as much of the background material as Tamir felt he could divulge without compromising any Mossad procedures. They walked silently to the hospital through the cold streets while light from the weak winter sun filtered through the clouds.

"This is where Hawthorne was beaten, you know," Lester said, "coming out of this very same station. I'll use that event to get Elizabeth to start talking."

"I suppose it's not a very safe neighborhood."

Lester shook his head vigorously. "Don't let the appearance fool you. It's actually quite safe. The police

151

couldn't remember a similar incident here in the last two or three years. Most mysterious.''

Tamir felt vindicated. His original instinct about the less-than-menacing nature of the area was correct. Not that it really mattered, of course. Or did it? he wondered. ''Most mysterious'' was an apt description.

Lester had arranged to do the questioning in his own office, with Tamir sitting in an adjoining room with an open transom that allowed him to follow the conversation. Lester began by indicating that he was concerned about Hawthorne's beating.

''But the police have already asked me questions, Prof. Lester. I told them all I knew.''

''I know that, but some new information has come to my attention and I thought it might be better if I talked about it with you—and no police about.''

''Oh dear,'' the woman said, clearly flustered.

''Now just relax, Elizabeth, and remember you are not under suspicion of anything.'' Lester reached into his jacket pocket and withdrew the picture Yossi had given Tamir at the embassy.

The woman's face turned red when the photograph of her and the Libyan was brought out.

''What's Ahmed got to do with this?'' she asked incredulously.

''Tell me about your friend Ahmed.''

Elizabeth was near tears. ''We met in Professor Hawthorne's office. He had come about some heart problem, but it turned out to be nothing serious. He was very friendly and brought me . . . '' She paused and said in almost a whisper . . . ''flowers. No one had brought me flowers for a long, long time.''

''Then you began going out with him?''

''Yes.''

''You became lovers?''

"Yes." She hung her head. "What has he done? Something really bad?"

Lester ignored the question. "Did he ask you about the work I was doing?"

"What do you mean?"

"My work for the military, on nerve gases." Tamir and Lester had decided on the train that there was no sense being too coy now.

Elizabeth looked genuinely puzzled.

"Ahmed never asked me anything about that."

"What did he ask you about me."

"Nothing."

Lester scowled at the secretary. "You'd best tell the truth, Elizabeth, it's the only way I can help you with the police."

"I am telling the truth," she said defiantly. "He was only interested in Professor Hawthorne's work."

"Hawthorne?" Lester said with surprise. In the next room Tamir also couldn't believe what the woman had just said.

"Yes. The new heart drug. He wondered if it would help his condition. Irregular heart beats."

"What drug is this?" Lester asked.

"It has no name, it's still experimental, just a number, 368. No wait, it has sort of a name, Frankfurt 368. Professor Hawthorne has been doing experiments with it. Very excited about it, he was. He was writing a paper about it when his arm got broken."

Tamir slapped his forehead. So that's what it meant, not part of an address or a flight number, but an experimental drug. He of all people should have guessed that!

Lester, only momentarily nonplussed, returned to his questioning. "What did Ahmed want to know about this drug 368?"

"He wanted to know if Professor Hawthorne had been talking to many people about the experiments he'd been doing and if the drug was becoming well known."

"Was it?"

"Not really. Professor Hawthorne was actually about to announce his findings in a medical journal. Until then, he didn't plan any news conferences or things like that. Ahmed seemed relieved when I told all this."

Realizing he had stumbled on something important, Lester pursued the line of questioning. "Did Ahmed say if anyone else wanted to know about 368?"

"No."

"Did he ever mention another name to you?"

"No."

"Did you hear him talking to other people on the telephone?"

Elizabeth wrinkled her brow. "There was only one person he spoke to from time to time. He never called him by name."

"What did he talk about?"

"I don't know, it was all in German."

Tamir let out a deep breath. Perhaps the puzzle was starting to come together. If he only had some time to match one piece with another . . .

"That's all I know, Professor Lester, really, can I go now?"

"Certainly, Elizabeth. You've been very helpful," Lester said.

"Will I ever see Ahmed again?"

"I think not."

She slumped in her chair. Lester felt sorry for her. She had been badly used.

When she had left, Tamir came in from the other room. "You were fantastic," he told Lester.

154

"Thank you. I must say, this is all still a bit of a mystery to me. I hope your chaps can figure it out. But it does seem clear that Hawthorne's beating was no accident—and to think, I was only using that incident as bait to get her talking! Who would ever have thought it would lead to Hawthorne's experiments? Look, I'm going to get Hawthorne to give me a chemical analysis of the drug he's studying. It's a heart medicine, isn't it? I've got a crazy hunch it may be related in some way to the compound that our chemical warfare people are so excited about. In fact, I wonder if . . . " Lester's words trailed off.

Tamir decided he would have to pursue that aspect later. Right now, it was time to talk to Shlomo again.

• • •

This time the telecommunications center in the embassy had little trouble finding Shlomo.

"You're to be congratulated for unraveling the enigma of Frankfurt 368," he said with obvious delight after Tamir had briefed him on Lester's conversation with Hawthorne's secretary. "But, tell me, why was the Libyan interested in Hawthorne's work? What does this drug have to do with the nerve gas?"

"I don't know. I'm as surprised as you are, but it's a good bet that Hawthorne was attacked by Libyans—or their hirelings—to keep him out of his laboratory and the public eye for awhile. The key seems to be the experimental drug he was working on. It may be related to a new kind of poisonous weapon even more dangerous than the standard nerve gases."

"Perhaps the company that makes 368 is somehow involved in helping the Libyans make their new chemical weapons."

"But why should a reputable drug company . . ."

"Who said they are reputable? You assume too much. Think critically."

Tamir remembered what Lester had said. "Libyan money could corrupt anybody."

"Maybe some people are already corrupt. It's also possible there are other things that we haven't even considered. There's much room for speculation."

"There's nothing more I can do here in London. Let me go to Frankfurt."

"Agreed. I think you've done well, but I've been told we civilians don't have much time left." Shlomo had to yell into the receiver because the connection suddenly became bad.

"How much time?" Tamir wanted to know.

"Two weeks."

"And then what?"

"It's out of my hands. It will be strictly military. Knock out the Rabta plant, hoping that's where the stuff is kept. That means lots of headlines, a bad press, and what's worse, casualties for us."

"Ground action?"

"Probably. At first the air force wanted to do it alone, but it sounds like we will also need a commando unit, especially after what Lester told you. If we don't do a precise job, we'll kill a lot of innocent people. If it comes to a ground assault, we'll need knowledgable people on site."

Tamir frowned. "Do you want me back for that?"

"Ideally, yes."

"How would you arrange it?"

"There's a regularly scheduled El Al flight every day from Frankfurt. I'd give you 24 hours notice to be on it. A seat would be held for you until takeoff."

"Fine, but how will you contact me in Germany?"

"Don't worry; that's my affair. Your job is to get over to Frankfurt and see what is going on there. Remember what we have to find out. How is 368 related to the new poison gas, or whatever it is? Does the same drug company also supply the Libyans with material for nerve gases? Do they have people working with the Libyans?"

"You expect me to do all this on my own?"

"Of course not. There are others working on it too. They'll help you. Take a morning flight to Frankfurt. Check in at the Grand Hotel. Messages will be sent to you there. Any questions?"

"Not for the moment," Tamir replied wearily.

Shlomo hung up, his face a mass of scowls. He glowered at Zvi. "Finally we're figuring out what's going on, thanks to Tamir. It shows you what a good choice I made. But how the hell do they expect me to do all this in two weeks?"

"One week, Shlomo. Your air force friend called a little while ago. The word came down from the Prime Minister's office to prepare for an air strike on the eighteenth."

Shlomo whistled. "One week!"

"The Prime Minister must know something we don't," Zvi concluded.

"He only knows what we tell him," Shlomo answered. "Yesterday he was told the Rabta plant has been reactivated but would not be able to produce any chemicals for at least two weeks. That's what our people in Libya informed us and they're usually reliable."

"This is the Egyptian connection? The ones who first alerted us about the German scientists at the plant?"

157

"Correct," Shlomo said. "Ever since we returned the Sinai to them, the Egyptians have been very helpful. Shows you that even sand has value."

Shlomo reached for the phone and spoke to a senior Mossad supervisor for several minutes. "The Prime Minister's convinced the Libyans were behind the Heathrow attack," he told Zvi, "and he's also convinced they're planning more trouble. They're completely irrational. He wants to give them a bloody nose as soon as possible. And number one on his list is the chemical warfare project. Apparently the Chinese have shipped some of their new, long-range ground-to-ground missiles to Tripoli via North Korea. They can be fitted for chemical warheads, according to the Prime Minister, and used against our cities. Of course, the Libyans would deny any involvement in a mass poisoning; they'd blame it on the Protectors of Islam or one of the other fanatic groups."

"Would the Libyans really do it?" Zvi asked incredulously. "Use poison gas against cities?"

"The Iraqis got away with it with the Kurds and Iranians, didn't they? A slap on the wrist, that's all their 'punishment' was. The Prime Minister's afraid the Libyans are crazy enough to try anything. He doesn't want to give them the chance to use their new toys. If he can scare them off, the Iraqis will get the message as well." Shlomo paused. "But I still want to see if we can do it my way! Use the military only as a last resort, and if we have to use them, keep casualties to a minimum with our superior intelligence data. Therefore, the more we can learn about the plant, the better. Hopefully our people in Germany can provide that."

"Capt. Tamir will be part of that team very soon," Zvi said.

Shlomo nodded. "And you too, Zvi, you'll be on that team too. After all, if you hadn't snatched the

158

Libyan in Cyprus we'd know a lot less than we do. Tomorrow you also will be on your way to an airport." He patted the younger man on the back. "Report to Tev in Frankfurt. Use our embassy contact in Bonn to locate him. Keep a protective eye on Tamir, he's still new to all this."

"And another protective eye on Dr. Goodman," Zvi added.

"Exactly. Good luck."

Frankfurt, January 13

Peter Werner's funeral service was held in the Lutheran church that his family had belonged to since the end of the war. There were personal friends, neighbors, workers and executives from the plant. Most of the mourners also went to the nearby cemetery for the short burial service. For the rest of the day Johanna and Goodman stayed with Trudi and the children while people trooped in and out of the Werner house to pay their respects. Goodman was relieved when night came and the depressing day came to an end.

In the morning he felt better, but the more he thought about that day's trip to Hannover and Bergen, the more he decided it might be a good idea to talk with Inspector Slitko again. After instructing Johanna to call Stern's office to make sure he would be available later that afternoon, Goodman drove the Porsche to the same police station he had visited after the holiday in the Harz was abruptly terminated by the news of Werner's death.

Slitko was at his desk, surrounded by piles of official-looking documents. He appeared harried. "These are busy times for me, I'm afraid. Because of the terrorist incident at the London airport, all police units throughout Germany have increased their vigilance on our Arab guests. We've already picked up some particularly dangerous characters in the East, but our manpower is overstretched. I'm going to be transferred to the airport detail in a few days, a really busy unit. But right now I'm still here and I remember promising you I would be available for any possible assistance, so please feel free to unburden yourself. Anyway, I need an excuse to get out from under the pile of paperwork." He pushed a report away and leaned back, his hands clasped behind his crew-cut head and a thin smile on his lips.

"I've been doing some unofficial snooping, Inspector. Initially, it was simply out of respect for Peter's memory and at the request of his wife and also his sister-in-law, whom I happen to be inordinately fond of. But as I've got deeper into my 'investigation,' as I like to jokingly refer to it, I seem to get more confused and more unsure whether the death was really accidental."

"Why?"

"For one thing, this idea of industrial espionage is certainly not as far outside the realm of possibility as I had first thought. Richter says he would not at all be surprised if the Communists were after his new drug. Whether Peter's death is somehow linked to such a scheme is conjectural, but since he did carry a pistol, he must have been afraid of something."

"What can I say? Anything is possible, but do you or Dr. Richter—or Mrs. Werner for that matter—have anything more substantial than theories for me to sink my teeth into?"

"No, not really," Goodman admitted. "In fact, the little hard evidence, if I can call it that, is to the contrary. With the assistance of a Danish colleague, I met with a leading East German research expert in my field."

"And?"

"He denied any involvement in a possible conspiracy to steal Frankfurt 368. But he is apparently interested in getting his hands on the drug for reasons that aren't clear."

"Who is he?"

"Grundig from Berlin. Does the name mean anything to you?"

"No." Slitko pushed the intercom button on his desk console. "Check into a Grundig from the

East," he said to his assistant. "What's his first name?" he asked Goodman.

"Ernst."

"Ernst Grundig," he repeated into the intercom. "Bring in whatever you have and check with the federal anti-terrorist unit headquarters. It's still in Bonn. They haven't moved to Berlin yet." Slitko released the button and looked up at Goodman. "What else did Grundig say?"

"Not much. Didn't seem to like Richter, though."

"Interesting that he knew of him."

"He seemed to know a lot about both Richter and his company. He even knew of Hans Meyer."

"Who?"

Goodman repeated what he remembered of the demonstration outside the restaurant and Meyer's subsequent flight the next day.

Slitko could make no further sense of it than did Goodman.

"I wonder if he's the one who's been feeding Grundig his information," the policeman said. "Anyone else you've met seem unreliable?"

"The weakest link in the research protocol appears to be Walter Stern, the cardiologist from Hannover. He has no background in this kind of work, yet Richter gave him the assignment, supposedly because of mutual friends. But Richter's so anti-Communist, I can't see Stern being an agent for the Russians or other cronies."

The policeman laughed. "Why not? It's a perfect cover. If your mind runs to that sort of intrigue, Stern would be a natural."

"So you think it's a possibility?"

"It's possible. Not probable, but possible. But in any case, what can I do to prove or disprove it? And

162

how does that tie in with Werner's death, if at all?" He pushed the intercom button again. "Add Walter Stern, physician in Hannover, to your check list." To Goodman he said, "For all we know, Stern could have come over from the East years ago as a plant before the Wall came down; that would be an interesting tidbit. I don't see what else I can do. I certainly have no manpower to stake out the plant. We're just too busy to allot much time or effort to accidents that may not have been accidents." Slitko grew somber. "But if anything comes of these checks, I'll let you know. Meanwhile, go ahead and see Stern and whoever else you have to. Find out about the pistol and so forth. And then come to me with something solid to work with. Otherwise, I'm a prisoner of this." He swept his arm across the mound of manila-folder reports. "Who else did you say you were seeing?"

"A headwaiter at a nightclub, the Nightingale."

"Oh, that place. One of our local dives that Turks and Arabs flock to for work. And plot in, and God knows what else. Maybe you'll uncover a *real* plot for me; now that would be an unexpected dividend of your investigation." He laughed. "Good luck, Doctor, for me it's back to work. If anything comes through on these security checks, I'll ring you up at the Werner home."

Johanna was waiting for him when he returned. "Stern will be in his office all afternoon," she said. "Let's get started. We can eat on the road."

The morning was bright and sunny as they drove north. Each took turns at the wheel of her Porsche after she had gone over their route on a map. The countryside was dominated by small villages and farms as they approached Hannover.

"I hope when we're through with Stern we can close the book on this subject," Goodman said

163

wistfully. "Maybe everything will fit nicely into place so we can tell Trudi that Peter's death *was* an accident."

"You'd like to tell her that," Johanna said, "but I can tell from your voice that you're not so sure."

"There *are* some odd-shaped pieces to the puzzle, that I have to admit, but since you're going to drive for a while, I think I'll take a snooze." He rested his head on the seat support and closed his eyes.

Sleep was long in coming. Too many thoughts were whirling around in his head. Instead of finding answers to questions, each question seemed to lead to a new question. It was bewildering. If he hadn't promised Trudi . . . His head lolled to one side as he finally dropped off. Johanna let him sleep until they reached the outskirts of Hannover. "Wake up sleepyhead, you've been napping for almost an hour." She left the autobahn and took one of the main highways into the heart of the city.

"What is this fellow Stern like?" Johanna asked as they approached the office address.

"He's sort of unctuous. I could feel Peter's dislike for him at the first day's meeting when he turned up late, and then again at the dinner in the restaurant when he turned up half drunk and with a streetwalker to boot. There's little doubt that it was Richter who persuaded Peter to take him on. From what Hoffman has told us, the reasons were more personal than professional, but I'd like to see what Stern has to say about it. And also, I'd like to go over his data."

"It's a shame that Peter's work had to be questioned because of Stern's negative results."

"There's no doubt about that. I was trying to be enthusiastic up to the time Stern presented his findings, but after that it was hard. Peter knew it, too. Of course, Eriksson kept up his interest, as had Hawthorne."

164

"Poor Hawthorne."

"Yes, 'poor fellow' is right. I hope he's recovering all right. It still wouldn't surprise me to learn his beating had something to do with 368."

When they arrived at Stern's office, they found several patients, but no doctor. "Dr. Stern is expected back from the hospital at any time," the receptionist said.

They sat in the waiting room, a very plush and ornate affair. They didn't have too long to wait. Stern, red faced as usual, came charging in, but pulled up short when he saw them. "Ah, this is a surprise. I thought you left for New York after Peter's funeral."

"No, not yet. I'm helping the Werner family. This is Peter's sister-in-law, Johanna Bauer. We'd just like a few minutes of your time."

Stern greeted her politely. "My condolences, fraulein. I'm sorry I couldn't attend the funeral. I had patients to see, which is the case again today, so if it can be brief . . . "

"It can," Goodman said.

"Then come into my office." He took off his coat and placed it in a closet. The office was as ornate as the waiting room. "Now, what can I do for you?"

"Apparently, you were one of the last people to see Peter alive."

"Yes, that could be."

"Anything unusual occur before you parted?"

"No, not really."

"But hadn't you and Peter had words on previous occasions?"

"Peter and I disagreed about the results of the 368 tests, or rather their implications. I don't think he believed that his drug could not work as wonderfully in people as it did in animals."

"Why, then, was the last meeting so much more amiable?"

"I take it you've been talking with Dieter Hoffman," Stern said, arranging papers on his desk. "Yes, we had a nice visit. I brought some sweets." Satisfied with his papers, Stern sat on the edge of the desk and resumed his conversation. "As I remember, the day before he died, he showed me reports that made it appear that some patients could really respond to the drug. I was very surprised because I didn't know there were any other data in humans. After seeing that, I agreed that perhaps I had been hasty in drawing too much inference from reports on my few patients. I returned the next day with a peace offering. That seemed to cheer him up. You know, he had been very depressed for several months. Perhaps because of family matters, I don't really know. Do you?"

Goodman ignored the question. "How did you get involved in this study in the first place?"

"Dr. Richter asked me."

"But you have little experience in clinical research."

"I have some," he said defensively. "Enough for the company to rely on my ability to recruit patients and maintain adequate records."

"You know, I never really examined the records of the patients you presented at the meeting. I wonder if you could show them to me."

"You want to see them now?"

"It will only take a minute."

"I really don't know if I can find them on such short notice."

"I think the company would appreciate any help you could give me. I know Peter Werner would."

"That's a rather crude way of putting it," Stern said coldly. "I'd appreciate your cooperation," Goodman said again.

Stern searched through a file cabinet behind him. "Here, see for yourself."

Goodman glanced at the electrocardiographic records. "These are the pretreatment tracings. What about the posttreatment ones?"

Stern went back to the file and came out with more. "All in order, I can assure you." He spoke smoothly but seemed to be tense.

Goodman looked through this folder a little more slowly. He compared the electrocardiograms of each patient before and after treatment. Every tracing contained runs of irregular beats similar to the control tracing. There was very little difference. To be sure, he again searched through the control tracings and compared them to the tracings done after the drug had been given.

"Do you think Peter's death was an accident, Dr. Stern?" Johanna asked while Goodman inspected the medical files.

"I think, miss, that your brother-in-law died of an accident. Is there any evidence to the contrary?"

"Some things are not as they should be," Johanna said matter-of-factly. "Perhaps the Communists are interested in the drug. That's not so crazy, is it?"

"Well . . . "

"Dr. Richter thinks it possible."

"Is that so?"

"That's what he told us," Johanna said.

"I have no reason to doubt Dr. Richter's judgment," Stern said, "but still . . . "

Goodman handed back the manila folders. "Thank you. You see, it only took a few minutes after all."

Stern smiled. "If everything is in order, I must return to my patients now. Excuse me."

"Not very helpful, was he?" Johanna said when they were back in the car.

"Actually, he was; the pre- and posttreatment electrocardiograms were so similar that I have serious doubts that he ever gave the patients the drug. I think he rigged the study to show that the drug was ineffective."

"Then it's almost as if he wanted the drug to fail! What a frightening thought! Why? And what can we do?"

"Nothing, for the moment. According to Grundig, Becker knows a lot of company dirt. Maybe he has some of the answers. We might as well continue on to Bergen before it gets dark."

Bergen, January 14

They drove in silence until they reached the outskirts of the small town of Bergen. Here, the nightclub waiter's directions proved useful and within a few minutes they had found Becker's farm. His compact house was dwarfed by the surrounding fields. Occasional patches of dirty snow left over from the most recent storm dotted the brown earth. Their knock on the door was answered by a pleasant-faced woman of middle age, rotund but not obese.

"We're here to see Herr Becker," Goodman explained.

She turned and called, "Heinz, someone to see you."

Becker came to the door. "Yes?"

"You probably don't remember us, but we were with Peter Werner the last time we visited the Nightingale. May we come in?"

"Peter Werner," Becker muttered, and looked toward the floor, shaking his head. "What a tragedy. Of course, come in."

The interior of the house was much more appealing than its exterior. A roaring fire lit up the hearth and the furnishings, though simple, were attractive and functional.

"Would you like some coffee?" Becker asked. "My wife makes an excellent pot." He went to the fireplace and added another log.

"Coffee would be fine," Goodman said. Becker signalled his wife to prepare the coffee and she hurried to the stove that occupied one of the corners of the large room.

"You are a long way from Frankfurt," Becker went on. "You have other business in this area, or are you on holiday?"

"A little of both," Johanna said.

"How did you know where I was?"

"One of your colleagues told us."

Becker seemed surprised. "They're usually very protective of my time off."

"We told them it had to do with Peter Werner."

"Yes, that name would help. He was a great benefactor to the club and especially to me."

His wife returned with the coffee and they arranged themselves around the fireplace. "I was quite upset when I heard of the accident," Becker said. "I planned then to take some time off and come up here. This is my refuge from the hustle and bustle. My wife and I come here often. We call it a farm, but it really just gives us enough vegetables for our table and that of our friends."

"You come here in the winter, too?" Johanna said in wonderment.

Becker's wife laughed. "Anything to get out of the city. We are country people."

"But why so far?" Johanna asked.

Becker's tone, which had been lively and friendly, now changed. His face grew heavy. His wife stopped laughing. "This area has other features besides its rich farm land that are important to us. But I don't think these things would be of interest to you."

"They might be if they're related to Peter Werner's death," Goodman said.

"What do you mean?"

"There's some concern that his death was not really an accident."

Becker seemed shaken. "Not an accident? That's hard to believe."

170

"Perhaps another company, or another country, was interested in the new drug he was working on. Stranger things have happened."

"You mean the East Germans, or the Russians, don't you?"

"Yes. Do you know some of their people?"

Becker smiled. "From a long time ago, but I haven't seem them for years. I have no idea whether or not they'd be involved in that sort of thing."

"You haven't seen them recently?"

"No."

Goodman was not about to give up. "Is there anything that you could tell us about your relationship with Peter Werner, or to his company when you worked there, that might at all be important?"

"I don't know how important it is, but I'll tell you what he did for me, if that's of any help. As for the company, I'd rather not go into that. Too many unpleasant memories."

"Peter told us you left to go into the nightclub business."

"He was being kind. I was fired by Richter, thrown out. I was lucky to get a job as a dishwasher until Peter helped me."

"What did Richter have against you? I thought you had worked for the company for quite a while, especially with Peter's father."

"I had, but when Peter's father died, I let some things off my chest that I had kept within me for some years out of deference to his father. Richter did not like what I had to say."

"Did Peter disapprove also?"

Becker looked away. "I didn't tell Peter these things either. It was better that he didn't know." He spoke softly. "Now that he's dead, I suppose it doesn't matter if you know." He looked at Johanna. "You're his sister-in-law, aren't you?"

"Yes."

"You'll have to decide if you want your sister to know."

Johanna looked puzzled, but Becker did not notice. He was staring straight ahead. "Maria, I'm going out for a while."

"You're taking them . . . there?" his wife asked.

"Why not? It's easier to talk about it. I'll get my coat."

Becker got into the back seat of the Porsche and directed Johanna out onto the road that led to the town center. But before they reached it, he had them turn off to a different road. They drove by a large British army base, part of a NATO contingent, then they pulled into a parking lot partially surrounded by a low brick wall on one side and a two-story flat building on another.

"This is the Belsen camp," Becker said in a low voice. "Also known as Bergen-Belsen."

"A concentration camp?" Goodman said incredulously.

"The remains of it. It was burned down in 1945. It is a memorial park now. Have you never been there, fraulein?"

"No," Johanna said softly, "but when I was younger, I visited Dachau in the south, the one that's kept intact."

"What has this to do with you and Peter Werner?" Goodman asked.

"Let's get out and walk about," Becker said, "and you will see. The building over there is the document center, but we can omit that for the time being. It's the grounds that I find more irresistible." He led them through a gate and into a sprawling veranda. A quarter of a mile ahead, a huge granite obelisk rose toward the blue sky. Stone paths led through the camp, or what had once been a camp. On the side of the paths were a dozen or so massive hedge rows, six feet in height and some thirty feet deep. Stone markers were implanted in the sides. Their message was simple: Here lie 1600 dead, April, 1945; here lie 2000 dead, April, 1945; here lie 1800 dead, April, 1945. Strewn about the grounds were a variety of markers. "Here lies my mother and father, and grandmother . . ." with names of people and camps. Not only Bergen-Belsen, but Auschwitz and Treblinka.

"They commemorate people who may have been here once, but died elsewhere," Becker said, as a scholar would in examining historical records.

There were also larger markers from Jewish committees and groups. Around each, small groups of people had gathered; children were present, though they were subdued. "The crowds are larger in the warmer months, but people come even in the winter. I think they find it quite a sobering experience." He walked ahead of them as if drawn by inner forces to the obelisk and its companion, a sixty-foot-long wall some twenty feet tall. At the base of the wall were floral wreaths from German, Jewish or Israeli groups, each with its inscription. "We shall not forget." "Never again." "To the memory of the martyrs." On the wall were inscribed—in a variety of languages—similar messages.

"This was not an extermination camp," Becker said, as they stared at the inscriptions in German, English, French, Hebrew, Polish, Dutch, "but, still, some 30,000 people died of torture, starvation, overwork."

"Is this where you lost your arm?" Johanna asked.

"Yes."

Goodman felt a great wave of sympathy towards Becker. He had always feared being in a place like this. Perhaps it was one of the reasons he avoided traveling in Germany. These were the kinds of sights that stirred dark thoughts in the recesses of his mind and emotions that, despite the years of assimilation, were never far below the surface: this was his heritage. As much as he might deny it, as much as he might wish it did not mean anything to him—an American scientist with no religious feeling—this was his heritage. And here was Becker, a survivor who obviously could not get the stench of the place out of his nostrils, who came here on his holiday when he could be many miles away. He had known there were Germans who were imprisoned—ministers, politicians, union leaders, any opponent of the regime, and here was one of them. He felt a great surge of compassion for this man.

"Isn't it painful to always be returning here?" he asked.

Becker nodded. "But I must do it."

Johanna protested. "You've suffered so much here," she said, reading Goodman's mind, "why not have your farm far away? Why always have this so close?"

"It must be close," Becker insisted. "I must never forget. Others do, but I must not, I cannot. It's God's will."

A chill wind began to blow. At the edge of the clear blue sky, a few clouds intruded, then more. "There will be snow tonight," Becker said. "It will cover the ground and the ghosts will walk again." He shuddered. "Follow me to the documentation center. The pictures are not pretty, but I must see them again because I am possessed by them." Mesmerized by his words, they followed him back to the entrance and into the low building. The exhibits depicted the history of the camp: savage tortures, official decrees, starving inmates, finally mass burials by the liberating British troops. When they could stand it no longer, he led them back to the car. "This is part of me," he said, "but not a good part. My story is not what you would expect, because I was not an inmate here." He paused while they stared at him in bewilderment. "I was a medical technician, a member of the SS. Yes, it's true, don't look so startled. And this is where I met Franz Richter. Yes, the same Franz Richter. He was the camp doctor. Wait, there's more. His assistant was Peter Werner's father." Becker stared straight ahead, while Goodman and Johanna recoiled in horror, withdrawing from Becker as if he were a leper. What he was saying about Richter and Peter's father was repulsive; it could not be true. Instinctively, they rejected it.

"You're crazy!" Johanna shouted. "Your time in this camp has unbalanced you. What nonsense!" Becker stared straight ahead, his face impassive. Tears streamed down Johanna's face. She beat her fists against the hood of her car until angry welts appeared on them. Goodman did not want to accept Becker's confession any more than Johanna could, but there was something in Becker's frozen stare that convinced him he was telling the truth. This was why Grundig had insisted he talk to him. Somehow, perhaps through former Communist inmates of the camp, he had learned about Becker . . . and Richter.

"Is that what lost you your job?" Goodman asked. "Your not being able to live with your past any longer?"

Becker shook his head. "No, I could live with my past. It's just that I could not condone it any longer. At first, after the war, I did. I still believed in the Party and the Führer. Then I became very religious—a born-again Christian like your former President Carter. And

175

what I did, or rather what I helped to do, in the SS was too much to accept without repentance. I had escaped being caught at the end of the war—that's when I hurt my arm, by the way, when a group of us fled south to Frankfurt—but the past gnawed away at me until finally I went to Richter one day and told him of my misgivings. That was bad enough, but what was worse was that I tried to convince him that God could forgive us if we admitted what we had done was morally abominable. Remember, he had hired me because I was one of his loyal orderlies, so he got furious and threw me out of the office. A few days later, he called me back. He was much more in control now. He said that I was through with his company and that if I ever told what I knew to the occupation authorities or the police, I would be killed. He said it just like that, and I believed him, because there was some gold and jewelry and important papers about veteran groups involved as well, and I was frightened. Peter's father had been dead a year by that time and he could not defend me, though I don't know if he would have if he could. Peter's father was not at all like his son. For him, authority was always to be obeyed. Always. That's when I contacted some of the survivors in the East that I had looked out for while they were in the camp. I was ready to bolt for the East, to live in an atheistic country, if necessary. Why not? God was within me, that's all that matters. Then things quieted down all of a sudden and I stayed.'' Becker stopped and glanced at the gathering clouds again. ''There will surely be snow tonight. The snow is worse here because it brings back memories of the inmates' bodies lying frozen where the guards had tossed them. I wish to go back to my farm now,'' he said suddenly.

They drove back in a much more subdued atmosphere. Becker's wife was waiting for him with hot

176

coffee, but he refused it, preferring to sit instead in front of the fire, his eyes fastened on the orange and blue flames.

"He's always like this after he goes there," she said. "Sometimes it takes days for him to return to himself. I tell him he doesn't have to go there anymore, but still he goes. The past is the past, I say. No one else cares, but he says he promised God, so who am I to quarrel with God?" She bustled about with the coffee, a huge apron wrapped around her waist.

"Peter never knew about his father," Johanna said, sipping from a mug.

"I never told him," Becker said, "and I'm sure Richter didn't."

"Peter thought they were in the army together, on the Russian front. God, if he had known the truth."

"Medical experiments, that's what they called it," Becker said impassively. "Inject substances directly into the heart, see how long it takes to die. They invented new kinds of poisons, chemicals that kill you instantly, paralyze your nerves. Records were kept, you see; Richter and his colleagues in the other camps even thought they might publish the results some day in a scientific journal."

Goodman felt as if he would vomit.

"He is a criminal," Johanna said. "He should be brought to trial. You could testify against him."

"He would have me killed. He's quite capable of it."

"There must be other witnesses . . . "

"All dead."

"The past is the past," Becker's wife said as she poured more coffee. "Why make trouble? For what? The troublemakers are all gone anyway."

"Who were the troublemakers?" Johanna asked.

"The Jews," Becker's wife said, cleaning up a spot of coffee that had dripped onto the table. "They're all gone now, so why stir up bad memories. If it hadn't been for the Jews, my husband could live a normal life instead of being possessed by demons." She shook her head in disgust.

"My wife is not as religious as I am," Becker said, "but we have been together some forty years. We have no children, only each other." He shrugged as if to say it is not she who is the pariah, it is I; it is not I who puts up with her prejudices, but she who tolerates my weird sense of values. "She wonders how I can mourn Jews when she believes our Fatherland was destroyed because of them."

The urge to throw up surged again, but Goodman fought it. Johanna looked as if she could bear no more either. It was all too much to assimilate so suddenly. He put the coffee cup down. "We must go now," he said to Becker, "but there is one question that bothers me. Knowing what you do about Richter, is it possible that in some way he could be responsible for Peter Werner's death? Could Walter Stern be?"

Becker smiled. "With Richter, anything is possible. But why would he want Peter dead? Peter was his chief researcher; he was always investigating new drugs. The one he was working on when he died could have made Frankfurt Pharmaceutica famous and Richter even richer than he is. Why would he want him dead?" He went to the fire and stoked it again. "Maybe Richter has also changed. After all, it was many years ago that he fired me. Maybe he's mellowed since then; maybe he doesn't make threats to kill people anymore. Stern? I don't know him at all." The flames flared up and Becker stood back. "Sometimes, I see faces in the fire," he said, "then I have nightmares. Tonight, with the snow, I will surely have them. But it is God's will. It is my penance. I accept it."

They left him staring into the fire, oblivious to all else, while his wife puttered about busily, still cursing the Jews for what they had done to her Fatherland.

• • •

Outside the Becker farmhouse, the first flurries were already falling; heavy, wet flakes that clung to Goodman's face and hair. From behind the closed door of the farmhouse, he could hear Becker's wife loudly berating him for sharing his past with a stranger. Finally, Becker told her to be quiet and the shouting ceased. An eerie silence descended on the farmhouse.

"I'm shivering," Johanna said, "but I don't know whether it's from the cold or from what I heard in the house and at the camp. To think of Peter's father in the SS with Richter—my God, it's too gruesome, too unbelievably gruesome." She shook her head, still not wanting to believe Becker's story.

"I feel the same way; I'm numb all over. Standing out here in the cold doesn't help. Let's get back in the car and try to make some sense out of it."

Goodman knew he could never forget the walk through the memorial park where Bergen-Belsen had stood, the mass burial sites, the inscriptions on the stones, the obelisk piercing the gray winter sky. For Johanna, it was an equally shattering experience.

"All this time, I thought of Richter and Peter's father as army doctors, doing their duty by saving lives and tending the wounded. To think of them in that camp performing those awful experiments . . . the SS were the worst of the lot and the most unrepentant. If Peter ever knew . . . "

"But he didn't," Goodman interrupted her. "He would never have been able to lie about something like

179

that, I'm sure of it. It would have gnawed at him to the point that he would have had to tell someone."

"I think you're right. His father never would have told him and I don't see why Richter would. Trudi never intimated that she knew anything about it. We're close enough so that she doesn't withhold things like that from me."

"Peter could never have worked for Richter if he had known. So in the final analysis, that's the best indication that Peter knew nothing of his father's past."

"I wish we had never begun this foolish investigation," Johanna said angrily, "it's brought us nothing but trouble."

Her remarks reminded Goodman of the reason for their trip to Becker's house, the investigation. What in the world did all this have to do—if anything—with Peter Werner's death? Becker knew little of Stern, and didn't think Richter had any reason to do in Werner. So what had they gained from the trip except the added aggravation of skeletons in the closet? Yet perhaps the news of Richter's notorious past should not be passed over lightly in this regard. Despite his mellowing—according to Becker—couldn't he still be capable of evil? What was this business about gold, jewelry and important papers? What could be so important about a veterans' group? Maybe it was wrong not to consider Richter as possibly being involved in Werner's death.

He posed the question to Johanna. "Do you think it's possible Richter could have had something to do with Peter's accident?"

"I'm still not able to even consider something as awful as that."

"I understand," Goodman said gently, "but right now the main question for us is to figure out if anything Becker told us has a bearing on Peter's death.

If Richter could possibly be involved, perhaps Stern is too. After all, he's Richter's protege in a sense."

Johanna shook her head. "I don't see the connection. Why would Richter want Peter dead? And if he did, how could he get Stern to arrange it?"

"I admit I'm fishing in the dark, but the way Becker spoke of Richter makes me think he's capable of anything. And the more I think about it, the more I have to wonder about Stern's role. After all, he did have arguments with Peter and he *was* the last person to be with him. Despite his air of bravado, I don't think he's a strong person. Maybe if we confront him with some of this material, he'll be more open-mouthed. To get back to Frankfurt, we have to go back through Hannover anyway. What do you say, is it worth it to see him again? Or have you had enough?"

"No, I'm all right, but do you think he'll want to see us again? He wasn't that crazy about our conversation this morning."

"There's only one way to find out. Put the car in gear and let's find a public phone."

They stopped at a petrol station on the outskirts of Belsen, and Goodman called Hannover.

Stern's tone was unpleasant. "What is it now? I'm busy with an office full of patients."

Goodman clenched his teeth, trying hard to control his distaste at dealing with Stern. "I've just had a very unnerving afternoon, Dr. Stern; in fact, a very unpleasant afternoon. Part of that unpleasantness was directly related to some rather frightening things that I've learned about your benefactor, Franz Richter."

"He's not my benefactor," Stern snorted, "he's a friend and colleague."

"Whatever you choose," Goodman said, "the information was still not good."

"What are you talking about? What sort of information was it?"

Goodman thought he detected a trace of concern behind Stern's gruffness. He took the plunge. Nothing ventured, nothing gained. "What I've learned makes me wonder if Dr. Richter had something to do with Peter Werner's death."

"Preposterous! Who's been filling your head with such poppycock?" The concern was now more blatant.

"That's not important. What is important is your friend's activities during the war. I think you know what I mean. His activities *and* those of Werner's father."

Stern gasped. His tone changed abruptly. "I can't talk about this on the telephone. Where are you?"

"Not far."

"Then come to my home later. There's no one there and we can talk more safely."

"Safely? Why do you say that?"

"If Werner were in danger from Richter, then I might be, too. I'll explain when I see you. I can't talk on the phone."

Can I trust him? Goodman thought, then wondered if he really had a choice. "Where's your home?"

"In Kaltenweide, a suburb south of Hannover. It's not hard to find. Just be at the town hall square at 1800 hours and I'll lead you there. My car is a tan Mercedes. Well, can you do it?" He paused, then said, "If there's a chance Richter can harm me, I'd better talk with someone about it and I don't want to go to the police."

Goodman glanced at his watch. He had almost an hour to get there. He could do it. "Yes, I'll be there."

"And the girl, bring her with you. She'll be very interested in what I have to say. I can't talk any more now." The phone clicked and there was nothing but dial tone.

Johanna was skeptical when he told her of Stern's suggestion. "Why should he think he's in danger? From what you've told me—and from what I've seen with my own eyes—he and Richter are great buddies. Paul, I don't see how we can trust him. And why does he want me to be there? You said he almost insisted on it."

"I don't know," Goodman admitted, "and I agree that he's not a particularly trustworthy character. We suspect he falsified the ECG reports and God knows what else he's done—but don't you see, Becker's confession may be the break we've been waiting for. It's stirred Stern up and made him awfully nervous—we've got to follow up on it. What do you say?"

She hesitated for a moment, then said, "Okay, I suppose you're right, but I have a bad feeling about it just the same."

As the Porsche headed south with Johanna at the wheel, the afternoon light faded rapidly and the snowfall became heavier. Near Hannover, the wind picked up and great swirls of snow covered the highway, forcing them to reduce their speed.

"This might turn out to be a real storm," Goodman said, "not like the light stuff we had in the Harz. Do you know any decent places to stay in Hannover? We might have to spend the night."

"We'll find something, don't worry. I'm going to take a bypass around the city center to get us to Kaltenweide a little sooner. How are we doing on time?"

Goodman checked his watch. "Twenty minutes to the hour."

"We'll just about make it."

While she devoted her attention to the highway, Goodman let his thoughts race ahead to the meeting with Stern. What lay behind that enigmatic exterior could be a Communist plant, as Inspector Slitko had intimated, or simply an inept "friend of a friend" that Richter was helping. Yet there could also be something more sinister. He turned over in his mind the various pieces of the puzzle beginning with Becker's soliloquy and running back to the strange session with the East Berliner Grundig and then to his meetings at the drug company. And, finally, to the accident itself, to the reason for Werner carrying a pistol, to the need for him to use a friend to do the clinical studies to prove the drug was of value. It all came back to the drug. What was the secret of Frankfurt 368? Why was it so important? A new—and appealing—thought jarred him and he sat bolt upright.

"What's the matter?" Johanna said worriedly.

"I've been toying with an idea, but it's all theory and no facts."

"I'm listening."

"Let's assume now that Trudi has been right all along, that somehow Peter's death was not an accident. Let's also assume that Inspector Slitko was right and the car was not tampered with. That still leaves the possibility that Peter was drugged in some way. And who was with him shortly before his death? Stern. A doctor with access to drugs."

"But the toxic screen was negative."

"Which only means the drug had to be one that is not routinely tested for, and one that can cause weird behavior—enough to lose control of the wheel."

"So?"

"Cardiac drugs can do that. Commonly used ones: digitalis, for example. In large enough doses it can cause fatal heart irregularities. The liquid preparation is easily obtained in your country. Stern could have poisoned Peter by slipping it into his tea. The whole business with the sweets could have just been a clever ploy. He eats one, Dieter Hoffman eats one, but only Peter dies, so it proves the sweets are harmless. No one would have thought of the tea.''

''Why would he do it?''

''Aha, that's the real mystery. That's what I've been wrestling with since Becker told us about Richter's past.''

Johanna was still puzzled. ''What does that have to do with Stern poisoning Peter?''

''We've been assuming that someone, Communists, perhaps, want the secret of 368 for their own purposes, or perhaps even to cripple Richter's company, but what we haven't considered is that Richter—or the people behind him—*don't* want the drug to be successful. He doesn't want his own firm receiving that much publicity. He has something to hide and he doesn't want a lot of attention directed at Frankfurt Pharmaceutica, or at himself. That would explain the attempts to prevent the drug from being evaluated properly.''

''You think that secret is Richter's background?''

''I think that's a good part of it, but there's got to be more. There are a lot of ex-SS men with similar stories. Maybe Stern knows what the rest is and he's scared enough to tell us. Working for Richter doesn't appear to be a whole lot of fun. Maybe Richter has something on Stern and he sees this as a chance to get back. Perhaps at last we're starting to get to the bottom of this thing.''

''It's also a little scary,'' Johanna said. ''Shouldn't we let Inspector Slitko in on what we've learned?''

Goodman reflected on that. She made sense, of course. They were exposing themselves to possible danger. ''Do you still have the card he gave you?''

185

She pulled off the road and fumbled in her pocketbook. Finally, she fished it out. "I hope he's available. Next petrol station I see, we'll pull into."

It was a while before they found one. Meanwhile, the snow on the highway was accumulating rapidly. Inspector Slitko was not in his office. His assistant explained that he had gone home with an upset stomach and was not expected back that day. Did Dr. Goodman want to leave a message?

"I'll try to reach him tomorrow. There's no message." After he hung up, he had second thoughts. No one knew where they were, but how was he going to explain the situation to anyone but Slitko? It was too complicated.

Johanna was as concerned as he was. "Do you think we should still go on?"

"We don't really have any choice, if we want to get to the bottom of all this."

As they drove on, the headlights of the Porsche illuminated the snowflakes, thicker and larger than before. Johanna took a sharp right and soon they were on the main street of Kaltenweide. It was barely dark, yet few people were about and many shops were already closed and shuttered. Despite the drifting snow, the main street presented no real driving problems . . . so far.

"This looks like more than just a little snowfall," Johanna said looking upset.

Goodman stared at the rapidly falling flakes. The sooner they heard what Stern had to say, the better he would feel. Being on the road on a night like this was not going to be fun.

"Paul, you've been so wonderful," she went on. "I can't thank you enough for listening to my sister's crazy ideas and staying to help us."

"We're not through this yet," Goodman said soberly. "Thank me when it's all over."

186

The windshield wipers rubbed noisily back and forth against ice as Johanna navigated her way through the snow. The street was open, but cars—the few that there were—moved slowly. Once they hit a slick spot and she had to work quickly to keep the Porsche from spinning off the road. "Damn him," she snapped, "why couldn't we have just met him in his office?"

"He sounded nervous," Goodman answered, "scared about something. If he has anything valuable to say, we've got to do it his way. I wonder how much further it is?"

"No more than a few streets, I hope. Sorry for getting angry. Maybe I'd better ask someone just to be sure." She stopped in front of one of the few shops still open. Directions from the owner placed them in the town square five minutes later, ten minutes after their rendezvous time. With the motor on they sat and waited. The windows became fogged from their breath. Goodman grinned half-heartedly. "If I wasn't so nervous, I could see this as a romantic setting."

"Why are you nervous? I thought I was the only one."

"It's the damn waiting. I hope he didn't have a change of heart."

They had only a few more minutes to wait. Fifteen minutes after the hour, a tan Mercedes sedan pulled up alongside them. The driver rolled down the window to show his face. "Follow me," Stern said, and rolled up the window again.

They started out slowly. The tan Mercedes made frequent turns. "I don't know whether he's taking a short cut or a long cut," Johanna said. After several more convolutions, she was convinced it was a longer version. "Why is he doing this?"

"To make sure no one's following us, I suppose." While the window of Stern's car was open, he had inspected the interior as best he could. The car seemed empty. The one thing they didn't need was Stern plus friends. A gnawing concern for their safety was intruding into his thoughts as he followed the progress of Stern's car. It was reinforced when they reached the outskirts of the town. The buildings became fewer and the distances between them longer. Finally, they were on an open road, thick with snow. There were no cars at all now. This was not the main highway they had used to enter the town, but a much narrower accessory road. There was barely enough room for two cars to pass each other. Despite the snow, the tan Mercedes picked up speed. The snow fell thick and heavy.

"He's turning *again*," Johanna said. "My God, this road's even narrower than the other." It was, in essence, a one-lane road.

"This must be the path to his house," Goodman said hopefully.

Navigating the narrow road was hard, but the Porsche held the road well as they crept along behind the Mercedes. Stern turned once more and now they could see a large house ahead of them with a circular driveway. The driveway passed right in front of the door of the house. An attached garage stood on one side. The headlights of the two cars played over the darkened walls and windows as they entered the driveway. The Mercedes stopped, the headlights still on.

"Well," Johanna said, "we're here. I must say I don't like it."

"Neither do I," Goodman said. 'Just on a hunch, sit here with the motor running while I go in with Stern."

"Why?"

'Just be prepared to pull out quickly if you have to. Think there's enough room for you to squeeze by his car?"

"Barely."

"And keep your headlights on his car while I get out."

"You're worried he's not alone?"

"Exactly."

"And if he isn't?"

"Then we're getting the hell out of here. I don't trust him much as it is, but if he has friends waiting for us . . . "

"Look, he's getting out of the car."

Stern had clambered out into the snow, turning off his lights. He moved toward the door of the house.

"Was there anyone in his car?" Johanna asked. "I wasn't paying much attention."

"No, I looked."

"And no one else seems to be here. The driveway is empty, the house dark."

"What about the garage?"

"I think it's a little in back of me. Should we check to see if there are fresh tracks going into it?"

"Can you do it quickly?" Stern opened the front door of his house, turned on the hall light, and stood silhouetted in the entranceway. He waved to them to come in. Instead, Johanna backed the car several feet so they could check the snow in front of the garage. She turned the car slightly so the headlights lit up the area.

"Oh, God," she said tersely. "There *are* tracks."

"They couldn't be from his car," Goodman said anxiously, "the snow would have covered them by now. I don't trust the bastard one inch."

Stern was shouting now, but they couldn't make out his words.

"Paul, it's not worth it, if there are others with him . . . I don't like this. Let's get out of here."

189

Something inside of Goodman told him she was right. "Okay, let's get back to Kaltenwiede and check in with Slitko again."

She gunned the motor and turned the car toward the tan Mercedes. Stern was still in the doorway, gesturing frantically. She missed the Mercedes by inches and roared out of the circular driveway. As they pulled out onto the narrow road, Goodman looked back. Another figure, then another, were now silhouetted in the doorway with Stern. A chill swept over him. "You were right," he told Johanna, "there are at least two more."

"I knew we shouldn't have come out here."

"We'll let Slitko and the police handle it. Just get us back to Kaltenweide. Do you want me to drive?"

"No. I'm all right," she said, but her voice was tremulous.

"Then watch where you're driving!"

She had barely missed leaving the road on a turn. The thick snow made visibility difficult, but in several minutes they were back on the larger road.

"Which way?" she asked.

"I'm sure it was left." He looked back. Two headlights were piercing the blackness behind them.

"Christ! I bet they're after us. Go the other way, drive north. They won't suspect that."

She turned right and roared down the road. The car was accelerating quickly despite the snow. When he looked back, the road was dark. Faster she pushed the engine, and still no one was behind them. They had probably turned left.

Goodman relaxed a bit, but his relief was premature. The car lost its traction and began to skid. Despite her manipulations with the steering wheel, accelerator and brake, Johanna couldn't control the skid. In a few seconds, they were off the road and deep

190

into a snow bank. Goodman jumped out and inspected the front wheels. "Stuck! Put it in reverse, I'll push." He pushed, but the wheel only dug in deeper. He tried digging away some of the snow with his hands, but still the wheels spun. Around him, the road was empty, the sky black except for the falling snow. On either side was dense forest. He went back inside.

"We're stuck, but good," he announced glumly.

"What do we do now?"

"We can sit here and wait for a passing ride, or we can walk to town and hope someone will give us a lift."

"But won't Stern turn around and come this way when he realizes we're not headed for Kalten-weide?"

She was right. "Yes. Maybe we're better off going in the other direction. We can walk along the road for awhile, then cut into the trees if we don't get a lift. There must be other houses around. As soon as we find one, we can call the local police. What do you think?"

"Let's try it," Johanna said.

She turned off the ignition and locked the car. They covered it with snow to make it less obvious from the road.

Buttoned up to the neck, heads down against the wind, caps pulled low over their ears, they began walking along the side of the road away from Kalten-weide, waiting for a car to come along. But there was nothing. After walking for ten minutes, Goodman stopped. "We can't take a chance on the road any longer. Any car coming from the direction of Kalten-weide might be Stern's. I say we take the next cutoff and get off this road."

"A few minutes more," Johanna pleaded. She cast a foreboding look in the direction of the trees.

191

They walked on. No cars appeared from either direction. When they came to the next side road, Johanna reluctantly followed Goodman.

"I don't like it any more than you do," he said. "I'm just a city boy. I'm cold and the snow looks deep to me. I wish we had the skis now. But what choice is there?"

Johanna nodded. "We're bound to come across a house. Okay, let's go on."

They turned down the side road with its deep-piled drifts. Around them the snow fell without let-up, the wind howled. The forest on either side looked dark and menacing. What the hell are we doing here, Goodman thought, but it was too late for second guesses now.

Frankfurt, January 13

The flight over the English Channel was one of the worst Tamir had ever taken, worse than he ever wanted to experience again. A constant battering from storm clouds left the British Air passengers nauseated and dizzy. No one dared eat anything. I'll take sailing on rough seas to this anyday, Tamir vowed. When the plane finally landed in Frankfurt, he was white-faced, sweating and still nauseated. As soon as he got through customs, he bathed his face repeatedly with cold water in the men's washroom. He knew it was not an auspicious way to start his German adventure. An airport bus took him to his hotel and he headed straight for the bed. Even though it was only one o'clock in the afternoon, he knew he was going to have to sleep this one off.

An hour later the telephone rang, awakening him. "Who is it?" he whispered.

"A friend of Shlomo's. I'm coming up."

Before he had a chance to reply, the person hung up. The voice was that of a woman. Strange, Tamir thought, Shlomo has never used a woman before. He had barely finished washing before a knock sounded at the door.

"Shalom. My name is Nora." She was dark-haired and pretty.

"Come in."

"How was your flight from London?"

"Awful."

"There's a bad storm moving all across western Europe. They expect snow here tonight."

"I didn't think anyone was going to contract me quite this soon."

She smiled. "There are new developments. Things are moving rapidly. Which is good for Shlomo.

He's under a lot of pressure to wrap things up before the military has to get involved. Shlomo doesn't like to see our work get into the headlines."

"How do you fit in?"

"My official title is assistant cultural attache at our embassy. This enables me to get around the country quite a bit. Right now, I'm going to be your liaison with some of our other operatives."

"Where are you from?"

"Haifa."

"I'm from Tel Aviv. I teach at the university."

Her smile returned. "I know all about you Capt. Tamir. Remember? I work for Shlomo."

Now it was Tamir's turn to grin. Her smile was very becoming, he decided. "How long have you been working for the Mossad?"

"I've been in Bonn for two years. My next assignment is ostensibly in the Foreign Office in June. I think its really for advanced weapons training. If you need someone to go sailing with this summer, give me a call."

Her frankness took him aback. "You needn't be bashful," she added, "if you don't call me in June, I'll call you. You're just as attractive in person as your dossier makes you out to be."

"If I seem surprised it's because you read my mind," he confessed. "We'll definitely go sailing."

Her smile disappeared. "Unfortunately, we have this business to finish first. And from what Shlomo told me, there are still some questions to answer."

"Some? I didn't know *any* were answered."

"We have reason to believe one of the Libyan connections, a German, is here now."

"What?"

"That's right. In Germany now. He left Libya two days ago on short notice. That's what our Egyptian friends tell us. We don't know exactly why."

194

"If he's coming to Frankfurt," Tamir said, "it must involve the 368 drug and the company that makes it. Something happened to make them panic. I'm sure that's why Hawthorne was beaten."

"There's been a death here too. Peter Werner, their head of research. There are some other things that I have to tell you. But I can do that in the car. We have to meet some other people. I also have a 9-mm Beretta for you in the car. There's a silencer that goes with it."

Tamir frowned.

"Shlomo warned you there might be trouble," Nora reminded him.

"At least you're not issuing me an Uzi."

"Don't be so sure. If we're not successful here, there are plans to attach you to a commando unit and drop you into the Libyan desert."

"Shlomo warned me about that, too. At the time it seemed preposterous. Now, I'm not so sure. All in all, I think I'd prefer to be sailing."

"Remember, we have a date in June."

"Yes," he said laughing, "but, first . . . "

"Yes." She glanced at her watch. "First we have to pick up Zvi—you know Zvi—then we have a long drive, so we better get started."

"Where are we going?"

"Hannover. We'll meet someone else there."

"I'm ready," Tamir replied, mystified at the complicated twists this assignment was taking and wondering who and what was waiting for him in Hannover.

PART 3

Defenders of the Fatherland

Kaltenweide, January 14

Goodman and Johanna walked until they didn't think they had the strength to walk further, but still there was no hint of houses, nor any traffic visible in the dim light. The road was totally deserted. Because of the increasing force of the wind, the snow had drifted into deep mounds and despite their precautions the snow was seeping into their boots, and their socks and feet were becoming wet and cold. As they continued, heads down against the wind, frost forming on their eyelids, they searched in vain for the telltale light of a safe refuge. But there was nothing. Johanna slipped twice, further adding to her discomfort. Her hands became numb beneath the gloves. Goodman fared little better.

When they had been walking for 30 minutes, they stopped and huddled against a tree.

"I'm getting very tired," she said, her voice almost lost in the wind.

"I've been tired for the last ten minutes."

"I'm wet, too."

"You worried about frostbite?"

"Should I be?"

"No, not yet. Let's keep going, motion is the best thing for cold legs."

The wind showed no sign of abating, nor did the snow as they trudged on. The road blended into the trees until it was difficult to distinguish the path from the surrounding forest.

"Look, there!" Johanna called suddenly, pointing ahead of them.

"I can't see anything."

"Look! Look! A building."

They walked faster now. The outline of a small building was silhouetted against the sky. The promise of relief from the storm gave them renewed energy. A low wooden fence separated the house from the road, but the snow had drifted over the top and it was no problem crossing it. They ran for the door and pounded on it. There were no lights inside. It seemed to be a small cabin. "No one's in there," Goodman yelled. "I'll try and force a window."

Goodman broke the only window and crawled in. The room was cluttered with gardening equipment from what he could make out in the dark. He unlocked the door for Johanna.

"What is this place?" she asked.

"I don't know, some kind of lean-to for a farmer or forester. What does it matter? It's better than being outside."

They found a canvas covering to drape over the broken window. "At least we're inside and out of the storm," he said. "Stern and his boys will never find us. We'll wait until morning or whenever it stops."

"Paul, are you sure we'll be safe?"

"Here, get those wet boots and socks off. Use these rags to keep your feet warm."

"Are you sure we're safe?" she persisted.

"As safe as we can be. If they are following us, let's just pray the snow obliterates our footprints."

They curled up against one of the walls. He looked at his watch. Eight o'clock. Was it only two hours ago that they met Stern in the town hall plaza? It seems like far more than two hours, he thought ruefully. "Here, rub your legs a while," he said to Johanna. "Does wonders for the circulation."

"What do you think Stern meant to do?" she asked, following his suggestion and feeling sensation return to her legs.

"He had a surprise planned for us. An unpleasant one, I'm sure of that. And I've a feeling that Richter set the thing up. We, like dopes, were all set to walk right into his little trap. Since we didn't, they'll be out looking for us tomorrow. Assuming the snow stops tonight, our best bet is to get hold of Slitko or the local police as soon as we can. How are your feet?"

"Much better."

"Good. Try and get some sleep. I'll keep my eyes open just in case we have visitors."

"And if we do?"

"We'll get those boots on and get out the way we came, but I doubt they'll have the stomach to be out tonight, either."

She leaned against him, snuggling closer for warmth. Outside, the storm howled on. Soon her breathing became steady and slow, and he knew she was asleep. He listened to her, to the hypnotic in-and-out of her exhausted breaths, while he wondered at the strange fate that had tossed him into the stormy night, running from people who intended to harm him, when he could have been in his cozy West Side apartment sipping a scotch and water and watching the traffic pile up on the approaches to the George Washington Bridge. It was such a paradox that he should have met in Germany this woman who meant so much to him. Lost in his reverie and overcome with fatigue, his eyes closed. He fought to keep them open. They closed again. Confused questions ran through his mind: Why had Stern set a trap for them? But was it really a trap, or had they overreacted when they saw the two other men? Maybe Stern needed bodyguards and had no desire to harm them. But then why had they taken off after the Porsche in such haste? There couldn't be anything so urgent to tell Goodman that it necessitated a chase through snow-filled roads. No, Goodman decided, it was implausible. It had to be a trap and they were right in fleeing. If they hadn't seen the tracks in the snow . . . He shuddered. It was time to stop playing detective and leave the rest to the police.

Somewhere outside the cabin a branch cracked. Goodman's eyes opened. Probably just a tree limb falling, but still . . . The illuminated dial on his watch read nearly 11 PM. He slipped Johanna off his shoulder and gently eased her to the floor, letting her sleep undisturbed. Moving slowly to the window, he cautiously raised the canvas flap. There was enough light to make out the scene in front of him. The figure of a man was making his way through the snow to the cabin, hands thrust in the pockets of his overcoat, cap pulled down over his face. In the darkness behind him, two yellow circles glared dimly. Stern's car, thought

Goodman immediately, and the man approaching the cabin is one of his henchmen coming to check it out. No farmer would be dressed like that.

Goodman looked back at Johanna sleeping on the floor. It would take the man no more than three or four minutes to traverse the field. Was there enough time for him to wake her, get their boots back on, and make a run for it? There was no other choice.

He shook Johanna. "Wake up," he said sharply.

Her eyes opened. "What's the matter?"

"We have company. Get your boots on right now."

While they tugged on their boots, he went over his plan: they would have to press deeper into the wooded area, hoping to stumble across an occupied building. They would need some lead time. He would have to neutralize Stern's henchmen, if necessary. His hands quickly felt around the floor for some type of weapon—a foot-long piece of steel pipe seemed to fill the bill. With the pipe in his hand, he positioned himself behind the door and had Johanna press herself against the wall near him. He gave her one squeeze for good luck, then transferred both hands to the pipe, prepared to raise it above his head as soon as the door opened.

He didn't have to wait long. The snap of another branch under the weight of a man's body announced the arrival of the visitor. The door opened with a quick burst of cold air and a flashlight beam shot against the far wall of the cabin. The beam travelled up to the ceiling, down to the floor, and then moved across the wall. Goodman heard the man's breathing, but he could see nothing beyond the flashlight beam. The man took a step into the cabin, but still the open door separated him from Goodman. The unmistakable odor of garlic assailed Goodman's nostrils. He raised the pipe above his head, feeling his hands trembling with the tension caused by the quick flow of adrenalin. Who was this man he was about to batter? Was he so sure that he was not some innocent motorist looking for refuge? The flashlight beam swung toward the near wall where they had tried to warm themselves and the man grunted in surprise. Small puddles of water were still present. Goodman felt the man stiffen. He knows we were here, he thought, and he's wondering if we're still here. He *must* be with Stern. The click of a pistol cocking settled the matter. It's now or never, thought Goodman, and he brought the pipe down on the man's head as he moved further into the cabin. Despite his lack of experience and his fear, Goodman's blow was almost perfect. Neither bone-crunching nor glancing, but effective: the man crumpled to the floor, flashlight dropping from one hand, pistol from the other. Relieved by the ease at which he had accomplished the feat, but still trembling, Goodman picked up the flashlight.

"I'm leaving the pistol," he told the terrified Johanna, "I wouldn't know how to use it anyway."

"Let's get out of here," Johanna said. Then she had second thoughts. "Maybe it would be better to leave the flashlight on. Stern will think everything's all right for a while."

She was right again, Goodman conceded. Instead of the light, he retrieved the pipe. That he might need again. The two of them bolted out the door and into the snow that was now boot-high. They ran in the opposite direction from the car's headlights, even though it meant entering thick woods. Branches cracked and snapped as they squeezed through the trees and denuded bushes. Half-running, half-walking, they tried to put as much distance between themselves and the cabin as possible. When Goodman stopped to look back, he could see a light in the cabin. So far, so good. They moved deeper into the woods, stopping only to catch their breath. Finally, the line of trees thinned out, and to their surprise and relief, they saw a light.

"Thank God," Johanna shouted, "my feet are freezing."

"So are mine," Goodman yelled back, aware that his toes were painful and tingling, early signs of frostbite. The trees ended abruptly at a wood fence. On the other side, across a cleared area, stood a house with light streaming from several windows. Smoke curled from a chimney. He helped Johanna over the fence, then scrambled across after her. The lit house was like a beacon and they ran across the field in anticipation of a warm and safe refuge. When they reached the house, they realized they had approached it from the rear. Quickly circling it, they found the front door. Off to one side was a garage and driveway leading to a small road. Fresh tire tracks were in the snow. Ringing the doorbell and banging on it accomplished nothing. No one answered.

"No car in the garage," Johanna said after a quick inspection. "They must have just left. What do we do now?"

Goodman pushed, kicked and shoved the thick wooden door, but it wouldn't budge.

"Break a window again," Johanna shouted.

The lower floor windows were surrounded by thick, snow-covered shrubbery. He couldn't get near them. Above the garage was a flat roof with windows leading to what must be a second-floor bedroom. That would have to be their route of entry. Extracting a ladder from the garage, he placed it against the garage and began to clamber up, but the base of the ladder didn't have firm enough traction. It moved and he fell off into the snow.

"You all right?" Johanna asked. He nodded and tried again. This time his foot slipped on an icy rung and he went sprawling again. Exhausted and bruised, he lay on the ground for several minutes to regain his strength. On the third attempt, he managed to make it halfway to the top when he heard Johanna yell out, "Thank God, the people are coming back."

He stopped and looked. A car was coming up the road and turning into the driveway. Salvation at last. It wasn't until the car was almost at the front door that he realized that it might not contain the inhabitants of the house. He kicked himself for having stood there watching it, wasting valuable minutes that he could have used to finish his climb, enter the house, call the police and later apologize to the owners and pay for any damages. He started to scramble up the remaining rungs.

A harsh voice called his name and a flashlight shone into his eyes, momentarily blinding him.

"Come down, Goodman, come down." Stern's voice came from somewhere behind that light. Johanna screamed. A slap in the face quieted her. There were three flashlights now, all pointing directly in his eyes. Goodman came down the ladder and a boot caught him in the chest. Gasping for breath, he slumped backwards.

"Now listen carefully, both of you," Stern said. "We have guns as well as flashlights, so do as you're told. Now stand up. Slowly. One at a time, Goodman first."

When Goodman stood up, his arms were pinioned behind his back and the hands tied securely. Johanna was next.

"Good," Stern said. "Now, follow me. We are going back to my car. You are not very good foxes, you two. You made it very easy for us to head you off."

Goodman tried to speak, but a fist caught him in the mouth.

"Later," Stern said. "You'll get your chance to speak later. Now come along quietly or my friends will be forced to break a bone or two. One already has a grudge to settle for his sore head. Move." He shoved Goodman forward. With Stern on one side and one of the thugs on the other, there was nothing he could do but comply. A sinking feeling in the pit of his stomach nauseated him, but he was still too stunned by the blows to fully comprehend their predicament or worry about it. Behind him, he heard Johanna cursing as she too was pulled along by one of Stern's henchmen.

"You stopped running too soon," Stern hissed. "The storm is ending and there is a whole group of houses just a hundred meters down the road. What a pity." He laughed again and pushed Goodman ahead to his waiting Mercedes.

Goodman should have been glad for the warmth of the car, but by this time he had recovered enough to feel only fear, all thoughts of creature comfort were now forgotten. In the front seat, Johanna sat pinned betweeen two guards, while Stern next to him delighted in making sure his elbow was constantly in his ribs. The Mercedes had little difficulty in retracing the route to Stern's house. By the time they came to the circular driveway, the snow had stopped. The moon was shining brightly. "What a fine night it has turned out to be," Stern said smugly.

Stern's house was now bathed in light. Another car was parked in the driveway, a black Audi. That must have been the one in the garage, Goodman thought. How stupid we were to ever come out here alone. "Out, please." The car had stopped. The five of them went into the house. Inside, the main room was fully lit. Stern shoved both of them into the center of the room.

He signalled his two henchmen, and they pushed Goodman and Johanna onto a sofa. Both of the thugs were swarthy muscular young men who were very attentive to Stern's directions. Red-faced and angry, Stern stood by the sofa. "You two have stumbled into something that you would have been much wiser to stay out of."

Regaining his composure, Goodman tried to brazen it out. "We didn't come here to be assaulted and tied up. We came here because we thought that you had something important to tell us, that you were in danger."

Stern chuckled. "My dear Dr. Goodman. What I told you over the phone had no semblance of truth whatsoever. It was simply a device to get you out here where we could have a more congenial atmosphere for our talks. My office in town was hardly appropriate to the issues you raised. And as for my phone being tapped or my life threatened, that was just a necessary ruse. You wonder why you were 'assaulted,' as you put it? Let us just say that my colleagues," here he indicated the two with him, "were very annoyed at having to traipse around in the midst of the snow trying to find you, only to be battered on the head. If you had come into this house when you first arrived instead of running away, we could have avoided many of these unpleasantries. Now let us get on with this." He signalled to his thugs and the captives' hands were unbound, their coats removed, and then the hands retied before they were pushed back onto the sofa. The room Goodman and Johanna were in was decorated in the ornate style Stern seemed to prefer, yet it was a comfortable room. In a different situation, Goodman could see it as a pleasant place to visit. But at the moment, the circumstances were not at all pleasant.

"What are you going to do with us?" Johanna asked.

"We shall see as we go along," Stern said. "We have some things to talk about." He sat opposite them and motioned for the two thugs to step back. "It would be very helpful for both of you if you were quite frank about the source of the information you told me about Dr. Richter."

"First, untie us completely," Goodman said.

"Then you'll try and run away again and we shall have a nasty scene. No, for the time being, let's stay as we are."

"What does it matter who told us?" Johanna asked.

"It matters to me," Stern said.

"And to Richter," Goodman added.

"Let's keep him out of this. I'm the one who's asking the questions."

Goodman snorted. "The first thing you did after we called was to get in touch with Richter. I'm sure that's why we're tied up right now. Whatever Richter wants done to us, you'll do."

Stern was not deterred. "I'm not going to argue with you. I just want to know how that bit of information came to your attention."

"Then you tell us what the hell is going on with Frankfurt Pharmaceutica and why it was so important that 368 be discredited?"

"Discredited? I thought the Communists were out to steal the drug, not discredit it."

"I've had second thoughts about that."

"Second thoughts? Hah! If you had never pursued the matter in the first place, you'd be back in New York by this time and not up to your neck in the mess you're in. Communists! I suppose you thought I was a Red spy planted in the company by the Russians!"

Goodman remembered the inquiry Inspector Slitko had begun into Stern's past. Was Stern merely bluffing or had that shot in the dark been true? The only way to find out was to keep Stern talking.

"So you deny any connection with the East?" Goodman asked.

"Of course," Stern retorted. "But let's not play games. I think you called me today for other reasons. Somewhere along the way your Communist plot theory got a bit worn out."

"Perhaps."

"Somewhere along the way," Stern mused, "someone said or did something to make the great Communist plot theory fizzle. So I ask you now," he said, leaning forward with his red face, "Why were you so sure the Communists aren't interested?"

"I spoke with Professor Grundig from Berlin."

"Aha."

Goodman took a deep breath. "He had no interest in the drug," he lied, "and he said no one he knew did."

"You believed him?"

"Yes."

"How did you manage to get hold of him so easily?"

"Eriksson helped. He trusts Grundig."

"Eriksson! That troublemaker! What nuisances you, Eriksson and Hawthorne turned out to be."

"Your thugs were the ones who gave Hawthorne the beating, weren't they?" Johanna said suddenly.

"These two? Ridiculous."

"Then two others."

"Perhaps."

"But you, yourself, killed Peter Werner."

"No, Miss Bauer, I didn't kill Werner, though I doubt you will believe me."

"You poisoned his tea," Johanna insisted.

"No, not at all. What an ingenious idea, but I'm afraid I never even considered it. But enough of that business. You still haven't told me how you heard the stories about Dr. Richter. Unless Grundig told you. Yes, that's very possible."

"Why are you so concerned about Richter," Johanna hissed. "He's a beast, an animal."

"He's nothing of the kind," Stern said heatedly. "He's a scientist with a great vision for his country, a defender of its greatest values. Someone is out to slander him and has duped you." Then, in a more subdued tone, "What do you know of Frankfurt Pharmaceutica, Dr. Goodman? What did Peter Werner tell you?"

"Very little. I could remember more if you untied my hands."

"Yes, I'm sure you would, but that won't be possible." He stood up and walked to the window, turned, and looked at them quizzically again, then went to the hallway with his "assistants."

"What will he do?" Johanna whispered.

"He needs the information about Becker for some reason. We've got to stall for as long as we can. And hope that the police will come."

"But how can they? They don't even know we're here."

Goodman had no answer for her.

Stern returned with the two thugs. "Mohammed and his brother will extract the information from you sooner or later. I suggest it be sooner for your sake. Arabs are notorious torturers."

The thugs tied Goodman to the end of the sofa, then turned their attention to Johanna. A few slaps in the face and she was sobbing hysterically. Blood trickled from her mouth. "They will beat her some more," Stern said impassively. "Then they will burn her breasts with cigarette butts. It's quite painful and disfiguring. Such a lovely bosom, too."

"I will say nothing," Johanna hissed.

"Oh, I know you won't," Stern said, "but your boyfriend will. He has a soft spot for you." He ripped off her sweater and blouse and methodically lifted the brassiere from her heaving breasts. The two thugs leered appreciatively while he calmly lit a cigarette. "When they're through with her breasts, they'll do the same to her vagina. Then they'll rape her. I don't think you want to see that, do you?" He handed the cigarette to one of them.

Goodman had enough. "Grundig told us to see someone at the Nightingale. He told us about Richter."

"Ah, the Nightingale. My friends here know the place very well." He draped Johanna's sweater around her shoulders. "We don't want her to catch cold, not yet anyway. So someone there confided in you. That could only be Becker. Becker has broken his vow of silence to our . . . organization. How unfortunate for him."

"What vow of silence and how do you know Becker? He doesn't know you."

"No, he doesn't, but I know all about him and his vow. He didn't tell you about that, did he? He's so preoccupied with God, he must have forgotten." Stern went into the kitchen. They could hear him dialing the telephone, but his words were mumbled. When he came back, his red face was beaming. "Excellent! A trip in the country for you two. I'm afraid I can't join you though. I have a little business with our friend Becker. I understand he's on his farm in Bergen. Now that the snow has stopped, it should be a lovely night for a drive."

"What about us?" Johanna said.

210

"You two? Why, you're going to be escorted back to your car. We found it by the road where you left it. Here, get dressed." He untied her hands and she put on the sweater and blouse. "Your coat too. We don't want you to be without your coat. And now you, Dr. Goodman." Goodman's hands were untied and his coat thrown at him. He put it on as slowly as he could.

"Good," Stern said. Quickly, their hands were retied.

"Why do we need our hands tied again?" Johanna protested.

"You don't," Stern laughed, "but we do. We don't want you giving us any trouble on your trip." He pushed them down on the sofa again. Looking at his watch, he told his thugs, "Start in about ten minutes." Turning to his two prisoners, he said, "Good night, Dr. Goodman, Miss Bauer. Have a pleasant trip. I don't think we shall meet again."

When he had left, Mohammed and his brother opened a bottle of brandy, but drank none of it. They talked to one another in their native tongue, which was unintelligible to Goodman.

"You're taking us somewhere, aren't you?" Johanna said nervously.

Mohammed smiled. He went to a cabinet and pulled out a bottle marked ether. "You're potential troublemakers for Dr. Stern," he said in heavily accented German. "Just like Becker is. So we're going to put you to sleep, bury you in the snow near your car and drench you with this brandy. Before you wake up, you'll freeze to death—two more drunken victims of the sudden storm."

"With our hands tied behind us?"

"We'll cut the ropes, but by then you'll no longer think about running off. You'll be frozen stiff!"

211

Desperately, Goodman tried his last ploy. "We're not the only ones who know about this business. Why don't you stop and think a minute before you're caught . . . "

Mohammed laughed. "No one will catch us."

"What about Becker's friends at the Nightingale? They'll want to avenge him if anything happens to him."

"They'll do nothing and if they do, we'll settle with them in time."

Mohammed opened the bottle of ether and began to empty it into a small towel.

"Why are you so loyal to Stern?" Goodman said quickly. "Is it money? I can get you more."

Mohammed laughed. "No one has more money than Stern and his friends. But there's more involved than money. Their enemies are our enemies. It's too bad you'll never find out about it." He shoved the ether-soaked towel into Johanna's face. She screamed and squirmed unsuccessfully to avoid the fumes. Goodman pulled again at his bonds, but they held fast. Johanna gave him one last imploring look, then slumped over. Mohammed poured more into the towel and advanced toward Goodman. There was no longer any way to stall him and no way to escape, but Goodman mustered up one last burst of energy and sprung to his feet. With a laugh, they pushed him back on the sofa and put the towel over his face. The sickly sweet smell of the ether enveloped Goodman and he felt himself growing dizzy. Then, as suddenly as the towel had been shoved at him, it was jerked back. Through misty eyes, Goodman saw Mohammed's face contorting wildly, his eyeballs rolling upward, blood spurting from his mouth and neck. He clutched at his head and fell backward; his brother plummeted heavily to the floor beside him. But the ether fumes had done

212

their job and Goodman could no longer keep his eyes open. The last thing he saw were three men with guns standing in the doorway of the living room. Two of them he didn't recognize, but one of them was a tall bearded man that Goodman knew he had seen before. The Romanian, he though incredulously, before sleep enveloped him. Ionescu, the Romanian. What was he doing here?

• • •

In his dream, Goodman was whirling, his arms and legs spread apart like the points of a star. The space that he was in was large and empty—he wasn't sure that it was outdoors—it could have been a vast, treeless field, or indoors—perhaps an airport hangar or convention hall. That much detail was beyond the recognition of the dreamer. What was important was that he was above the ground and whirling and that he was a target for Stern, who stood somewhere beneath aiming a rifle at him and periodically firing, laughing all the while, a high-pitched evil whine of a laugh. From the vantage point of the dreamer, Goodman was able to "see" himself being punctured by the bullets, painless projectiles that drew no blood but left small perfectly round holes where they entered and exited. When his body had been hit by a dozen or so bullets, it started its descent to the ground, much like a wounded but still viable bird. Instead of whirling, it fell in a gradual trajectory, closer and closer to the ground, where Stern stood laughing. Only it was no longer Stern; it was Mohammed, the Arab, and the laughter was deeper and even more menacing. The sound reverberated through the vast chamber. Then the faces were gone, and instead, Goodman saw where his falling body would land: a large silver-blue lake with a perfectly clear and ripple-free surface. His body struck the surface of the lake with a whack. The water was ice-cold. The body plummeted like a stone through the water until it hit the soft muddy bottom. He thrashed about, futilely trying to prevent the mud from covering him completely, but he couldn't move his hands or feet—they were paralyzed. Harder and harder he pushed and pulled, trying to make them move, but the mud level gradually rose until it was about to obliterate his vision.

That was when he awoke, straining against the ropes that bound his hands, conscious of cold water dripping down from his forehead into his neck and shirt. Opening his eyes revealed images that were still blurred and misty. He wasn't sure of where he was, but he knew he was alive and *not* sinking in some quagmire. Gradually, he recognized that he was still on a sofa in the living room of Stern's home. At the other end of the sofa Johanna was staring at him through half-opened eyes, still fighting off the effects of the ether. Why cold water was dripping from his forehead he had no idea, but that was only one of the questions he wished he knew the answers for.

He turned his head slowly, not sure where Stern's men were, afraid that any sudden move might alert them. Then he saw the one called Mohammed sprawled face down on the floor where he had fallen, arms and legs akimbo. Goodman couldn't see the body of Mohammed's brother; it must have fallen behind a sofa or chair. All of a sudden, he remembered those last few seconds of consciousness before the ether took effect. The guns that were used must have had silencers, for he had heard no shots. Where was the Romanian, the mysterious Romanian? Where were the men with him? Who were they? Goodman's eyes now took in the entire living room. There were no other persons in it besides himself and Johanna, half-awake on the sofa, and the Arab dead on the floor. In the back of the room, a grandfather clock ticked noisily. Other than the ticking, the house seemed deathly still. As he strained his ears, the sound of muffled voices could be heard coming from another room.

Johanna stirred again; both eyes were now completely open. She, too, looked about her. Her expression on seeing Mohammed's body was one of obvious relief. "What happened?" she whispered incredulously.

"You wouldn't believe it," Goodman replied slowly, "but our mysterious Romanian shot both of the Arabs. I don't know why and I don't know where the other body is. I just hope we're not going from the frying pan to the fire."

"It can't be any worse than what Stern and his friends had planned."

As if on cue, a voice came from the direction of the kitchen. "I'm delighted you two are awake," he said in English. The accent was familiar to Goodman, but he couldn't yet place it. "I'll be with you in a minute, I'm just finishing a snack." A chair scraped bare floor, a refrigerator door slammed, footsteps approached the living room as Johanna and Goodman exchanged bewildered glances.

From the hallway, Tamir entered the living room, a reassuring smile on his face. He felt exhilarated by the part he had played in the rescue mission. Nora had briefed him about Goodman and 368 on the drive from Frankfurt to Hannover. At first he was annoyed that Shlomo had not confided in him, but, as Nora said, that's the way he operated: keep sources of information separate so that data could be confirmed independently. In Hannover, the rendezvous with Tev and Zvi had gone smoothly, but from that point on, it was a wild drive through the snow to Kaltenweide and the shootout with the Arabs.

"Now that you're awake, I can untie you. Otherwise, I was afraid when you came to you might think me one of them," Tamir said, gesturing toward the body on the floor, "and get violent. So, please excuse the ropes, but, believe me, I have your best interests at heart." He smiled again and pulled out a switch blade knife from his pants pocket. The blade sprung out like a scythe as he advanced toward the sofa. Seeing the two of them instinctively stiffen, he stopped to reassure them. "As this one and his brother were about to drag you outside towards a very frigid death, we shot them," he explained. "It seemed the quickest and most reliable way of keeping them from harming you, which I must repeat, is one of the things that I am most concerned with."

"You'll excuse our skepticism," Goodman said tersely, "but after what we've been through, we're taking such statements with a few grains of salt."

"Try and bear with me. The ether had pretty well knocked you out, but when it started to wear off, I applied some ice packs to your head. That and a whiff of smelling salts speeded things along quite a bit. Then, while you were waking up, I had a chance to get something to eat. I must admit, my eating habits have been very erratic since I first became involved in this business, Dr. Goodman."

"It hasn't helped my digestion either," a similarly accented voice announced the arrival of the bearded Romanian. "Following you and Miss Bauer about for the last ten days or so has probably cost me several enjoyable inches here." The Romanian patted his flat abdomen. "But I think you'll admit it was worth it." While he chuckled, Tamir cut the ropes from Goodman's and Johanna's feet, then went to work on their hands.

"I'm confused," Goodman said to the Romanian. "I saw you at the ski resort. You looked familiar, but I didn't realize you had been following me. I'm still not sure what you're doing here beside the obvious fact that you saved our lives. I know you did that because I saw it before I passed out. But what I don't understand—besides how you got here in the first place, since following us today couldn't have been very simple—what I don't understand is why a Romanian engineer is interested in our safety?"

"So many questions," the Romanian said, smiling. "I'll begin by saying that my friends and I have a certain interest in the drug called 368, but actually your safety is of more concern at the moment."

"Is that why you've been following us?" Goodman asked as the final bonds were cut, "to protect us?"

"One of the reasons. Why did you think I was Romanian?"

"We checked your passport at Herzberg."

"Admirable. But you can call me Tev—and forget that Romanian stuff. No more Anton Ionescu."

Johanna stood and stretched. "That feels good."

"Some brandy?" Tamir asked. "I've noticed Stern has a good assortment. While you're drinking, I'll tell you a little story."

While the four sat and sipped brandy, Tev pulled a pipe from his jacket pocket, stuffed it with tobacco, and puffed away contentedly. The sound of voices in another part of the house grew louder, but Tev assured them with a wave of his hand that the situation was under control.

"Don't worry about that," said Tamir. "It will be clear in a minute. Our third colleague, Zvi, is interrogating the surviving gunman. Incidentally, my name is Dan. My colleagues have been interested in you, Dr. Goodman, since you arrived in Frankfurt some two weeks ago. Or, more precisely, while you were in transit. We are all fascinated to learn more about your dealings with Frankfurt Pharmaceutica because we're very concerned about that firm and some of its infamous personnel. Once it became apparent to my colleagues that your relationship with the company was benign, they viewed you in a different light, a more sympathetic light. When Peter Werner died, they worried very much about your safety. Thanks to the hotel clerk at the Fürstenwald, Tev had a good idea where to look for you on your skiing holiday."

"But why are you both so interested in Frankfurt Pharmaceutica?"

Tamir looked at Tev, who stopped puffing on his pipe and responded.

"To begin with, let us say we have been puzzled by some aspects of the company for awhile. And then, there was this business with the new drug. Was it really a cardiac drug or wasn't it? What was your involvement with it? It became necessary to have a closer look. I knew of your travel plans and took the liberty of examining your manual the first night you arrived. If you felt like you were drugged . . . "

"The beer! I remember the taste of the beer was not quite right."

"Exactly. A hundred marks for the desk clerk and I have several minutes of uninterrupted picture taking."

"But why not wait until I was out of the room?"

"That was my original plan. But at the time, I felt it was important to see the material before the Monday meeting, and when your plane was delayed and you didn't want to go out walking, as the hotel clerk tried to suggest, which, other than sleeping, is what most people do after a long flight, I had to improvise."

"And after that, you kept an eye on me."

218

"Yes. As I said, we were afraid for your safety, especially after Hawthorne was beaten. Werner's death we didn't anticipate, quite frankly. But knowing Richter and his crowd, we shouldn't have been too surprised."

"You know all about Richter, then?"

"We know a little," Tamir said, "and we want to know more."

"Who exactly are you two?" Goodman asked in exasperation.

"You're an intelligent person. Knowing what you do about Frankfurt Pharmaceutica, your first guess would probably be a German security agency, or perhaps even the CIA, but you'd be wrong. We're Israelis."

"Knowing what I do about Richter and seeing how efficiently you disposed of the two Arabs, I'm not surprised. But, what you want with 368, I can't imagine."

"Israelis!" Johanna said. "If you're Israelis, you must be after Richter!"

Tamir smiled. "Close enough. We're after him, some of his friends, some important files of his, and a certain Libyan connection. Since we've just shot two men who were about to kill the two of you, I feel you're entitled to know a little about it. If you want to know, that is."

"What are these files?" Johanna asked.

"They are lists of his cronies from his SS days" Tamir answered, "an invaluable list of names and addresses of suspected war criminals living all over the world. They are called the Defenders of the Fatherland; they are the most fanatic of the old Nazis and behind-the-scene supporters of the current neo-Nazis."

"So *that's* the veterans group Becker was talking about," Goodman said. "But what's the connection to Frankfurt Pharmaceutica?"

219

"Richter apparently bankrolls many of them through the company" Tev continued, "a sort of pension fund for the older ones without steady incomes. It's a unique arrangement but one that keeps their lips sealed. What caught the eye of our people in Vienna after Richter was identified, was that the company doesn't appear that prosperous based on its limited product line, but obviously was quite successful."

"They had stolen gold and jewelry from camp inmates?"

"True, but even that could only go so far. Now we understand where the extra income comes from. Some of these scientists have been supplying technical help to certain oil-rich Arab countries interested in acquiring chemical warfare capabilities. Unfortunately, we think Israel is the target, especially now that peace with the PLO seems a reality. We believe Richter has also set up dummy corporations in the Far East to ship raw materials and machines to these countries in return for fat fees. But this is something Dan knows more about."

Tamir sat down next to Goodman and put a reassuring hand on the cardiologist's shoulder. "The work of Tev's group in Europe has been mainly directed at Nazi hunting, but my role is different. I'm a neuropharmacologist at Tel Aviv University who was doing my reserve stint in the army when the Mossad selected me to help them unravel a chemical warfare threat they had stumbled into. It turned out the nerve gas I was checking out was not the real story, another compound was." He unfolded the piece of paper with the chemical formulas on it. "Does this one look familiar? A scientist in London found out it looks amazingly like the compound Prof. Hawthorne was evaluating, the one you came to Germany to learn about."

Goodman studied the formula. "It does resemble Frankfurt 368, but I notice substitution of sulfhydryl groups for hydroxyl ions in several places."

"What do you make of that? Could that transform a useful heart medicine into a poison?"

Goodman stroked his chin, rubbing the stubble of the day's growth. "There is a well-known effect in my field called 'pro-arrhythmia.' Drugs that are designed to prevent or treat irregular heart beats instead precipitate them, often with fatal consequences. I suspect the compound you're concerned about is such a drug. Whether it preceded 368 or followed its discovery I can't say. But since Richter and Peter Werner's father worked on poison compounds in the concentration camps, I have a hunch that 368 is actually a 'safe' derivative of the toxic compound. For all we know, Peter's father saw the possibilities for such an anti-arrhythmic drug and never told his son its origins."

Tamir interrupted. "And the original compound is the one that Richter now is selling to Libya and Iraq and God knows who else?"

"Exactly. That explains why he didn't want 368 to be successful. Don't you see, the company really didn't need the money and it certainly didn't want the publicity it would have received. The press would have loved to do stories about a small company finding a wonder drug and once they started snooping . . . "

"Of course," said Tamir. "Once they started doing background stories on Richter, it would have opened a Pandora's box. They had to discredit their own drugs. That's why they had Hawthorne beaten. What about Peter Werner's death? Do you think that's part of it? We know you've been busy checking every possible clue."

"To tell you the truth, I don't know, but I wouldn't put it past him."

221

Tamir stood up. "Our job now is simple—but very difficult. We have to bring an end to the whole business—stop Richter's dealings with the Arabs and either dispose of or uncover those SS criminals that are still alive. But we need your help. Think before you answer. There may be more danger involved."

Goodman and Johanna exchanged glances. More danger? They were lucky they hadn't been killed already.

The voices in the other part of the house grew louder. Tamir saw the quizzical looks on their faces. "My other colleague," he said by way of explanation. "Come, I'll show you."

He led them to the rear of the house, to one of its bedrooms. Propped on several pillows was Mohammed's brother, barely alive, his shirt blood stained, his arm hanging limply at his side. Goodman saw what looked to be a morphine syringe on the bed near him. Sitting on the other side of the bed, talking to him loudly in Arabic—while the pain-killer did its work— was a handsome, blonde-haired man about Goodman's age. The Arab was talking freely. The Israeli looked up briefly when they entered the room, then returned to his questions.

"He's trying to find out more about the Libyan poison gas factory," Tamir explained. "Stern's bodyguards may or may not know about it. If this one does, Zvi will find out in a few minutes." Zvi knew the Arab was bleeding internally and he didn't have much time left to finish the questioning. But he nearly had all that he needed and he pressed on.

The Arab was telling what he knew. The injection that had been given to him, a combination pain killer and central nervous system stimulant, had done what no amount of physical coercion could have. The

222

modified "truth serum" had eased the pain of the abdominal wound and loosened the doomed man's tongue as well. A little extra shouting by Zvi did the rest.

"Stern told us that he would have a friend coming from our country," the wounded Arab said slowly, "an important man in the war against the Zionists. We would have to guard him carefully when he arrived."

"When would that be?"

"It was already. Last night he came. We took him directly to Richter. He gave Richter some papers. They talked a long time."

"What do you know about the papers?"

"Richter was happy to get them. He said 'Now that we have these plans they'll always need us,' or something like that. The other man laughed and that's when they told us to leave the room."

"Where is that man now?"

"I don't know. We took him to the airport this morning."

There was no more to learn from the dying Arab. His eyes suddenly rolled back and his breathing ended after a series of fitful rasping sounds.

"He's dead?" Tamir asked from the doorway.

"Yes." Zvi stood up. "He and his brother were sent here to work for Richter several years ago when he started having dealings with the Libyan government. They were more or less on call for whatever dirty jobs came along."

"Your 'truth serum' did a remarkable job, under the circumstances," Tamir said.

"Shlomo doesn't realize what a bargain he has in me," Zvi said, only half in jest.

"What did you learn about the poison gas expert?"

"He's already met with Richter. He's gone. If we're lucky, the man gave Richter the new plans for the Rabta plant."

"Who is this man you've been talking about?" Johanna asked, puzzled. "Is he a German?"

Zvi hesitated.

"She's risking her life for us," Tev reminded him, "there's nothing she can't be told. The same for Dr. Goodman."

"I have to hear that from Shlomo, Tev."

"It's okay, Zvi," Tamir said. "I'll take responsibility for this. Miss Bauer," he said, turning to Johanna, "this man that we're referring to is apparently an old crony of Richter's. He and some of his associates are experts on chemical warfare and a key link to Libya's chemical warfare schemes. We would have loved to talk to him, to learn all we could about their new plant that he helps run in the Libyan desert. But we were too late. We think that plant poses an immediate and direct threat to Israel, thanks to the latest in ground-to-ground missiles that they've acquired."

"I appreciate your candor," Johanna said, "but I'm not happy to see Germans involved in this . . . this barbarism."

"If you want to help us, you've got to go back to Frankfurt," Tamir said. "I'm sure Richter is so well organized that he has a large cache of documents, including all the files for the Fatherland organization and possibly the plant structure as well."

"Don't forget Stern," Goodman said. "He went up to Bergen to kill Becker, the SS informer, and probably the best witness to the medical experiments at Bergen-Belsen."

"Too bad!" Tev pounded his fist into a table. "If only I'd known, I could have tried to stop him. But I was thinking only of getting you two out alive. I'm afraid it's too late now to help Becker. Our main hope is to get the information from Richter."

"What about Stern?" Johanna asked.

"I'll take care of that bastard," Zvi said ominously. "The rest of you go to Frankfurt."

"There's a police inspector, Slitko, who's been very helpful," Johanna said. "Can we go to him?"

"I don't know how he'd take to some of our activities here," Tamir said. "It might be better not to confide in him too much."

"I don't understand why Richter can't be arrested and tried as a war criminal," Johanna said.

"The trials are often a farce," Tev explained. "The witnesses are old and easily browbeaten by defense attorneys. No, that wouldn't accomplish any useful purpose."

"But wouldn't the bad publicity ruin his career, his business?"

Tev smiled. "We can't count on that. Sometimes in these trials it's the aged war criminals who get the sympathy. But I grant you that in the case of Richter, it might stir up some interest if his connection with a network of these people were to be revealed. And at the least, the pension arrangement would cease. His business dealings with the Libyans would also embarrass the federal government. So if all else fails, that might be an acceptable fallback position. But first, we must try our best to get those files."

"What led you to Richter in the first place?" Johanna asked.

"We have contacts in the Nazi documentation center in Vienna. During Otto Strumpf's trial, Richter's name came up as a character witness. That's when a little digging proved useful. We kept our eye on him just on general principles. That's how we learned about his shipments to the Middle East and about his monthly checks to his colleagues around the world. But the 368 business confused us and for the last two weeks we've been trying to put pieces of the puzzle into place. It hasn't been easy, but you two turned out to be a big help.''

"Why don't you just kill all of these Nazi criminals?" Johanna said bitterly. "They deserve it. Like Eichmann did."

"The local authorities get upset," Zvi answered. "Or else we would do just that. We'll have to, tonight. When Stern returns we'll make it appear he shot two Arab intruders and was himself fatally wounded in the exchange."

"At this point, I don't give a damn what you do to Stern," Goodman said, "but don't underestimate him like we did. By the time you see him, he will have murdered Becker. He's a slimy creature and he's vicious; he'll do anything he has to for his sacred 'cause.' "

"I appreciate your concern," Zvi said, "but I think we can handle it. Now we have to get you two out of here before he gets back."

"We can't go too far without a car," Johanna said, "and my car is stuck in a snow bank."

"I have a chain in my car," Tev said. "Do you think a little tow job would do the trick?"

"Probably," Goodman said. "It's worth a try, but do you have enough time to prepare for Stern?"

"If you leave right now, we will. Finish your brandy and get your coats on."

Tev had parked his car on the road that led to Stern's driveway, but at a sufficient distance beyond the driveway entrance so that it could not be seen from the house. Now that the snow had stopped, he had little trouble driving back to the spot where the red Porsche had spun off the road. He attached the chain from the rear fender of his rented BMW to the rear fender of the Porsche. With Zvi's help, he was able to exert a gentle traction that succeeded in freeing the Porsche. Johanna turned the motor on without difficulty.

"How can we thank you for saving our lives?" she said to the Israelis. "Words are useless."

"Remember that we're going to need some help if we're to get our hands on Richter's files," Tamir replied. "You two are invaluable because Richter thinks he can intimidate you. My advice would be to drive back to Frankfurt and pay your police friend a visit in the morning. As far as he is concerned, you left this place early in the evening. You know nothing about shooting, etc. I doubt it will surface for a day or two anyway. Just tell him what you learned about Richter's past. Let him handle it from there. Don't mention Richter's friend or the files."

"And when will we see you again?"

"One of us will contact you at the Werner house in a day or so. Take care."

With a good-bye wave, Goodman and Johanna started off.

Frankfurt, January 15

Their trip back was uneventful. With the first light of dawn, they reached the outskirts of Frankfurt. By the time they reached the Werner house, the sun was up. "When do we contact Inspector Slitko?" Johanna asked sleepily.

"In a few hours, after we've slept a bit. Or maybe this afternoon."

"You do want to help the Israelis, don't you? You didn't seem too enthusiastic."

Goodman tried to explain his reluctance. "This has turned out to be a very dangerous adventure. It's no longer simply Peter's death. We've gotten involved with people like Richter and Stern who think nothing of killing anyone who gets in their way. Sure, I'm a little hesitant."

"But Paul, these are evil people. We have to do what we can to help. But I shouldn't be having to tell *you* this." Her tone was sharp. "I'm really surprised to hear you talk like that."

"Let's get some sleep" he said, ending the discussion.

It was early in the afternoon before they saw Slitko. They had to wait at the station house while he closeted himself with aides, but he seemed genuinely pleased when he greeted them in the hallway.

"Well, the amateur investigators! I understand you tried to reach me yesterday when my digestive tract was running wild. But fill me in on what you've learned."

"The most important thing we learned concerns Richter's past; how that ties in to Peter Werner's death is not clear."

Slitko seemed interested. "Go on."

"Richter, and Werner's father, were in the SS together. They were camp doctors at Bergen-Belsen."

Slitko's eyes widened. "That is interesting."

"They apparently conducted experiments on the inmates."

Johanna interrupted. "In simple terms, they were war criminals."

"But how did you learn this?"

"We were told this by a man named Becker who used to work with them," Goodman said. "We saw him yesterday in Bergen and then saw Stern in Hannover. Stern thought it was ridiculous, but there's no question in my mind that Stern purposely ruined the 368 experiments."

"Why?"

"Probably on Richter's orders. I don't think he wanted 368 to succeed, frankly. I don't think he wanted a lot of publicity because of his past."

Slitko scratched his head. "How are we going to tie this in with the car accident? Is it related?"

Johanna interrupted again. "Even if it isn't, surely Richter can be indicted for his past crimes."

Slitko wasn't so sure. "I must contact the special prosecutor's office in Bonn with this information and see what he says. And, of course, we shall contact Becker. Where can I find him?"

"He works at the Nightingale; he should be back from his holiday next week." Goodman felt uneasy lying, but . . .

"We can't wait that long. Where did you say he is in Bergen? I'll have the local police talk to him. As for Stern, we can wait a bit, try not to scare him off."

"What about Richter?"

"For the time being—nothing. If we get too obvious, he might decide to avoid an indictment by leaving the country. By the way, how long will you be here, Dr. Goodman?"

Goodman looked at Johanna. "As long as necessary."

"Good. You'll be staying where?"

"At the Werner house."

"I'll get back to you as soon as I know something."

Goodman and Johanna spent the rest of the afternoon in the downtown area trying to forget for the moment their experience of the night before. They revisited the photography galleries, stopped for Viennese-style coffee, and arm in arm, strolled through the crowded streets. The snowfall had little effect on Frankfurt and the shoppers were out in force.

"This winter will really be gloomy without Peter," Johanna reflected. "I hope you'll stay longer."

"I can't," he said, "I'm due back soon, but why don't *you* come to New York in the spring? Let *me* entertain you, and without mysteries and intrigues."

"Do you really want me to? Romancing a German may not seem so appealing to you when you're back in your own country!"

"I really want you to come."

She kissed him in front of surprised passers-by. "Then I think I will."

When they returned to the Werner home, Trudi was frantic. "A policemen called several times. Inspector Slitko. I didn't know where you were. It's about someone you know—Becker. Who is this Becker?"

"It's complicated, Trudi," Johanna said. "What was the message?"

"To call this number." She handed her sister a piece of paper. Johanna looked at Goodman. "They must have found Becker at his farm."

Goodman nodded. "Oh well, we might as well get it over with." He rang the number and talked with Slitko's assistant. When he hung up, he was smiling. "Becker's still alive. They've flown him here to the Central Hospital in a helicopter. Slitko's been questioning him. Stern must have botched the job again."

"What does this mean?" asked the puzzled Trudi.

"It may be important," her sister said excitedly, "but we can't explain."

"Slitko's office said we're to go to the hospital as soon as we can," Goodman added, grabbing the car keys. "And that's now as far as I'm concerned. We'll tell you all about it later, Trudi, but it's good news for a change."

● ● ●

The intensive care unit of the Central Hospital was a crowded ward crammed with electronic gadgetry and monitoring equipment. For Goodman, it was all too much like a busman's holiday; the setting was very familiar. For Johanna, it was a glimpse into a new and frightening world that she decided she could easily do without. The sight of patients attached to intravenous tubing, artificial respirators and numerous other medical devices was overwhelming. "My God, I hope I don't throw up," she said as they followed Inspector Slitko down the row of cubicles to the one that Becker was in. Slitko told her to sit down and compose herself while he and Goodman went to Becker's bedside. The former SS orderly was fully awake, but with labored, sonorous breathing.

"Stern's bullet severed his spine, and passed through his intestines," Slitko said in a low voice. "It paralyzed him, sent his blood pressure plummeting and made his pulse beat extremely faint. Stern probably assumed he was dead. He did better with his wife, though; killed her outright. Becker saw the whole thing. I'm afraid he won't live long."

"May I speak to him?"

"Of course."

Goodman leaned over the bed. Becker's eyes fixed on him, but there was no hint of recognition.

"I saw you yesterday. You took me to the camp." Still Becker's eyes showed no recognition. He leaned closer so that Slitko could not hear what he was going to say. "If you want to get back at Stern and Richter, you must tell me where the list is. You know what I'm talking about. The files on all the Defenders of the Fatherland. Do you know where they are?"

Becker's lips moved, but the words were inaudible.

"What's that you're saying to him?" Slitko asked sharply.

"I'm telling him how sorry I was for what happened to him."

"What did he say to you?"

"I couldn't hear."

"Let me try." Slitko pushed closer to the bed. Becker's lips moved again. The inspector put his ear toward Becker's face. "No," the policeman responded, "you can't have any, it will make you sicker. He says 'water.' He must want water," he explained to Goodman, "but he's to have only intravenous fluids. His insides are pretty well chopped up."

The effort of speaking seemed to drain the remainder of Becker's strength. He lay back, exhausted. His eyes closed and his breathing became even more labored. A nurse hurried them outside the cubicle where Johanna was still sitting. Though still pale, she was feeling better. "How is he?" she asked.

"Dying," Goodman said.

"At least he identified Stern for the Bergen police," Slitko added, "that will help. We've notified the Hannover police to bring Stern in if they can find him. I have a feeling he may have fled the country."

"What about Richter?"

"I think you and I had better pay him a visit in the morning. That is, if you're up to it. He may deny everything, of course. Becker was the only real witness to the charges. And as for Werner's death . . . " He shrugged.

"I'm sure he's involved in it," Johanna said.

"Perhaps, but proving it will be difficult if we don't catch Stern. I shall have a driver pick you up at nine in the morning."

When they were back in the Werner house, Goodman told Johanna of his aborted talk with Becker. "If he knew about the documents, he wasn't going to tell me, or he couldn't. I don't see how Tev will ever get them from Richter."

"Wait till Richter hears about Stern's death. That might unnerve him enough to give us a chance to be of some help to the Israelis."

"Help? How can we help?"

"You heard what Dan said."

"Look, they're especially trained for what they do. God knows how many PLO gunmen they've already disposed of the same way they disposed of Stern and his thugs. But we're just amateurs, and bumbling ones at that. We came damned close to getting ourselves killed last night and I'm not particularly eager to try it again."

Johanna looked at him strangely. "I wish Dan and Tev would get in touch with us. I want to help them no matter what you do. If these men had something to do with Peter's death—directly or indirectly—then they have to be punished—one way or another. The

Israelis seem to have a good idea of that: an eye for an eye. That makes sense to me."

"You're talking like a child."

"Oh, I'm a child, am I? Was I a child last night when those animals tortured me? When I trekked through the snow with you? You're the one who's acting like a child, Paul Goodman. My God, these men are murderers of your people. Here's a chance to get back at them. How can you refuse?"

"I'm not refusing," Goodman replied, his voice rising. "I'm just saying I want to think about it a little more before I endanger our lives again."

"Don't shout! You'll wake the whole house. And don't speak *for* me! I'm old enough to make decisions about my own life."

Goodman threw up his hands in frustration. "That's the whole point. You don't see the danger. You just rush in because you're a good guy and they're the bad guys. Well, I care for you a little bit more than you care for yourself, apparently. I don't want to see you dead." He glared at her and she glared back.

"I appreciate your concern, Paul," she said in a softer tone, "and I appreciate your staying on to help me find out more about Peter's death. But in some things I don't see eye to eye with you."

Goodman kissed her gently on the forehead. "I'm sorry for getting angry. I've grown sort of attached to you."

She put her head on his chest and he wrapped his arms around her protectively.

"I've grown sort of attached to you too," she said, "that's why I don't like to argue with you. But that conversation with Becker affected me deeply. Here I am reading and studying about the Hitler era—all very abstract and in the past—and all of a sudden my sister's father-in-law turns out to be one of the war criminals

234

that I detest. To go from there to facing death at the hands of Richter's gang was the final touch. The past is very much alive for me now, Paul. It's not any sterile dissertation. Whatever Richter had to do with last night, I want to know if what Becker said was true and if there is a list and a scheme to pension them off, as Dan says, then I want to see it come crashing down. It would be the most important thing I've ever done, more important than finishing my thesis in Heidelberg."

Goodman knew there was no point in arguing with that.

"Israel must be an interesting country," Johanna said. "I'd like to go there some day. Have you ever been there?"

"No," he admitted. "It's one of those things I've always said I would do then never got around to doing."

She looked up at him in a way designed to melt any remnants of anger that might remain. "Let's not argue any more. Let's wait and see what Dan has to say."

"I can't argue with you for very long," he said gently, "over anything. I care for you too much."

She lifted her face and they kissed each other hard and hungrily. Without a word, they undressed quickly, as they had done during their first time together in his hotel room. The pent-up emotions of their mini-spat were soon dissipated in frenetic love making; Goodman knew that they had created something between them that could not easily be undone. As she lay sleeping in his arms afterwards, he wondered again at the strange twist of fate that had thrown him together with a German girl who had more interest in Israel than he did.

The police car that picked up Goodman at 9 in the morning contained a very puzzled Inspector Slitko.

"We found Stern last night, or rather the Kaltenweide police did. He was dead, apparently killed in his home by two intruders whom he must have surprised and shot at. Arabs. It appears he killed them and was mortally wounded himself. The problem is the bullets that killed the Arabs didn't come from the gun Stern used to kill Becker and his wife."

"Becker's dead?"

"Yes, died a few hours ago. Very confusing case." Slitko drummed with his fingers on the window pane. "I'm afraid the reports are over the radio by this time, which means that Richter will be prepared. Too bad. Now we lose the element of surprise. But who would have guessed last night that Stern would be dead also?"

Slitko was right. Richter showed no surprise when they were ushered into his office. "You're here about Stern, I presume? Awful business. Thieves are everywhere, I suppose, but still, to shoot a man in his own home."

"Like the Werner case, there is still a great deal we don't know," Slitko said.

"What's this got to do with Werner, Inspector?"

"Oh, come now, Dr. Richter. Here are two doctors working on a new experimental drug for your company and you think it purely a coincidence they are now both dead?"

"I see Dr. Goodman's been putting ideas into your head. Maybe he's right, Stern was always jealous of Peter. Perhaps it's his fault Peter's dead, but what can we do about it now?"

"Stern was your protege," Goodman said. "If he was involved in Peter's death, then maybe you were, too. Perhaps you even had Stern killed to silence him."

"Poppycock."

236

"You see, Dr. Richter," Slitko said, "there are other allegations about you that raise questions as to your veracity."

"What sort of allegations?"

"Your wartime record."

"My wartime record?" The general director roared. "I was an army doctor, nothing more, nothing less."

"A fellow by the name of Becker says differently."

"Becker! A malcontent, troublemaker. I fired him years ago. He'd say anything to get me in trouble. Let him stand before me and accuse me," he said confidently. "I dare him."

"He's dead," said Slitko.

"Oh? Well, then . . . "

"But I have a feeling you knew that already."

Richter grunted. "You are too quick to involve me in your fantasies. First, you lie to me about a Communist plot, now you impugn my war record. If you have charges to make, do so in a court of law. But I warn you, I have powerful friends in the courts and in your own police department too, I might add. So be very careful that you do not make wild charges and libel me," he snarled at the policemen. "In that case, you will lose your job. And as for you, Dr. Goodman, my advice to you still stands: Go home. Gentlemen, good day. Back to work for me." He rang for his secretary.

Slitko had Goodman drive back to the Werner home. He was apologetic about Richter's insinuations concerning his influence in the Frankfurt courts and police. "It's probably true," Slitko conceded. "His contacts are numerous, as are his friend Strumpf's. That whole generation is not apt to get excited about war crimes, they would much rather sweep the whole thing under the rug since so many of their friends are involved. Unless we get some airtight evidence against him, I don't think we can get an indictment."

"What about the Federal prosecutor's office? Don't they have a large file on these things?"

"Yes, that's a possibility and now they have access to East German records as well. Yes, I'll pass on your suspicions to the special prosecutor's office. They weren't much help with the checks we did on Stern and Grundig, but maybe they have more on Richter."

"What about Werner's death? Can we implicate him in that?"

"You heard Richter. If there was foul play, it was all Stern's fault and Stern is dead. I must say, it's going to be even more difficult to implicate him in it."

"I notice you didn't mention Stern shot Becker, yet he seemed to know it."

"Yes, that was interesting. I believe he did know about it; he may well have ordered Stern to do it—Stern was his willing agent in a lot of things—but how can I prove it? I am somewhat bothered by why they did all these things. True, the publicity about the drug would have led to revelations about his past, but with all his connections, the war crimes trial might have dragged on for years without him going to jail if, indeed, he was ever indicted. So it's still a puzzle to me why it's so. But there's little more to do for either of us. I suppose you'll be going home now.

"Yes, soon."

Johanna was not the only one waiting for him. Tamir and Tev had arrived. Fortunately, Trudi and the children were out. Goodman reasoned the less they knew about the Israelis, the better.

Tev was puffing away on a pipe, the tobacco aroma a pleasant blend of apples and cherries. "I've

238

been regaling Johanna with stories about Israel. Perhaps she might take a trip there one day. Who knows?"

"Richter blamed Stern for everything bad," Goodman said, ignoring Tev's gentle barbs. "And the police have nowhere left to go. Slitko has some questions about things—if he knew of the list, that would answer his questions—but I don't think he's too eager to pursue the investigation any further. He sort of intimated that when he found the bullets that killed the Arabs didn't come from Stern's pistol."

"That was my fault," said Tev. "Hopefully it won't come back to haunt us."

"Forget about that now," Tamir exclaimed. "Our main job is to get your hands on the list." Turning to Goodman, he said, "Johanna tells me you had a chance to question Becker before he died. Any luck?"

"No, none at all. He just wanted some water."

"Well, we can always break into Richter's office and look around. Tev's quite good at that. You two game?"

"I am," said Johanna. "What about Zvi?"

"He's communicating with Tel Aviv. He'll meet us later. We'll have to wait until dark to go to the plant. No one will be there except perhaps a night watchman, whom we can avoid or put to 'sleep.' Gently, of course." Tamir smiled while Tev blew smoke clouds at the ceiling.

For a few minutes no one talked, then Johanna asked quietly.

"Tev, was your family always from Israel?"

"No. Romania. I was born there after the war and spent enough time there to speak the language fluently. That's why I use 'Ionescu' as a cover."

"You come to avenge," Johanna said.

239

"Yes, I come to avenge," Tev replied unabashedly. "Most of my father's and mother's families were murdered by the Nazis."

"And so you kill when you have to."

"Yes. When I have to. Come to Israel, visit the Yad Vashem Memorial to the Holocaust, then you'll understand."

"I will, one day," she said, and Goodman knew it was more for his ears than Tev's.

They had dinner at a nearby cafe. The Israelis made it clear that what they would be doing that night would be important, but it also would be risky. "After what you went through in Kaltenweide, you might have second thoughts about helping us," Tamir said. "I understand that. But it's important, you both know that."

Johanna didn't need any convincing. "I want to go with you. I don't care about the risk."

"And you, Paul?"

"I'll go."

"Good. We each have a gun, but neither of you have to carry one. I have a flashlight and some other equipment in the car. Shall we get started?"

• • •

Frankfurt, January 16

The Frankfurt Pharmaceutica plant was dark, the industrial park quiet. An eerie silence—no cars, no animals, no people. They parked at a distance from the main gate, leaving Johanna behind with instructions to slowly circle the area while Tamir, Tev and Goodman cut through the back greens toward the main building. Tev opened a back door, and they quickly made their way to the main office suite. He picked another lock, this one to Richter's office and, with the flashlight to guide him, found the file cabinets that he was looking for. While Tamir and Goodman kept watch, he went through one drawer after another. It was a tedious job and an unrewarding one. He then turned his attention to Richter's desk. It likewise revealed nothing of interest. They looked for a wall safe, but could find none.

"Zero," Tev said with disgust. "I'm surprised. I don't think he would would keep this stuff at his home. I thought for sure it would be here in his office."

"Maybe there's some place else in the plant he kept it?" Goodman asked. "Some hidden niche."

"But where do we start?" Tamir wondered, "Paul, pick your memory. What did Richter ever say about this place that would suggest a place that no one else knew about?"

"He never said anything, to me that is."

"What about Peter Werner? Did he ever say anything?"

"No, not that I recall."

Tev tried another tack. "Did Werner ever say anything about the plant that got Richter annoyed? Something special that Richter didn't want others to be privy to?"

"No. Say wait a minute." Goodman tried to recall the conversation on his first day at the plant. "At lunch, we were discussing the beautiful grounds and Peter remembered how he had first seen the place when he was little, how his father had told him they had built the plant here because of the scenery, the quietness, and so on. Richter got very upset, said it had nothing to do with that at all, that it was just to get away from the rubble of downtown Frankfurt."

"Hmm." Tev thought for a moment. "Did he mention any particular aspect of the scenery?"

"The pond. How pretty it was. How his father had liked it. That seemed to annoy Richter also. And once when I went out to admire the pond, I was ordered away."

Tamir banged his hand on his forehead. "Of course! How dumb of me. Come on!"

"Hey, where to?"

Tamir smiled. "Remember Becker's words to you?"

"He didn't say anything, all he wanted was something to drink."

"No, I'll bet he didn't say he wanted something to drink. I'll bet he just said 'water,' right?"

"Yes, but . . . "

"Water! The *pond* is water! Come on."

They retraced their steps through the empty halls, down the stairs, and out the back door. The night was clear with a full moon. With quick strides, they walked across the grass toward the pond. It lay in quiet beauty, the branches of the large oaks hovering over the stone bridge, the benches, and the still water like a protective mother.

Tamir flashed the light around the edges. "Look for some place where the grass had been disturbed or there's a large rock or anything out of the ordinary."

Their eyes fastened to the ground, they walked the perimeter of the pond. Nothing. They stepped back five feet and widened the circle. Nothing. Again at ten feet, then fifteen feet. Nothing. Tamir cursed. "Where the hell could it be?"

"*Under* the water, maybe?" Goodman suggested.

"No, it would have to be someplace that Richter could get at readily and without arousing suspicion." He flashed the light around the pond again. "Is there a bench or a sitting area? No, I don't see any. Then his light flashed on an oak tree. Maybe there." The base of the oak was examined. Nothing unusual. He flashed the light up the trunk, first one side then the other.

"Well, well," he said triumphantly, "look at this." On the far side of the trunk—the side away from the buildings—the bark had been removed and a patch of black tar inserted.

"Just a patch for a rotten bough that was removed," Goodman said. "What's so unusual about that?"

"The tar patch just happens to be at the level of a man's arms," Tev said. "Look around the edges for a lever." With the light on the tar patch, he felt the rim and the center for something to pull. But there was nothing. Goodman was right. It was a tar patch pure and simple.

Totally frustrated, Tamir stamped his foot into the frozen ground in disgust. "Think, Paul, what have I left out? You've been here before. Think!" He flashed his light around the grounds of the pond once more: trees, shrubs, bare ground. Nothing seemed out of order.

Goodman followed the beam of light. There was something missing, but he couldn't quite put his finger on it. "Flash it on the pond itself," he told Tamir.

The light moved across the water, over the stone bridge, back to the water.

The bridge, Goodman thought, what was there unusual about that bridge? Then he remembered. "The one time I was actually here—a security guard chased me off that bridge. Something about the structure not being sufficiently strong enough to support much weight."

"A stone bridge not being able to support a man's weight? That doesn't sound right."

"But, it gave the guard a pretext to order people off of it."

"*Exactly*. Whose orders was he following? Richter's no doubt?"

"Correct."

"I think we'd better take a look at the bridge." Tamir crawled its entire length on his stomach, feeling each stone, prying at its corners, probing for an opening. Oblivious to the cold, he continued his inspection on the side walls when the bottom support seemed intact to his careful scrutiny. He had gone only a third of the way along one side wall when he found the loose stone he had been searching for. He tugged at it briefly before it swung free, apparently suspended by some lever type of attachment.

Tamir flashed the light in the dark recess where the stone had been. There were several shelf-like partitions carved out of the interior of the back wall. "Someone's used this to store something, that's for sure," the Israeli concluded, "but whatever he kept here is gone now."

"You mean the files?"

"And other things as well," Tev replied. "Probably the entire Fatherland files plus concentration camp records, experiments, whatever an ex-SS officer would want to have records of, but not in places easily

244

found or identified with him. Very few keep their museums in their cellars anymore."

"Now that's interesting, because when I spoke to Becker, he made some reference to there being gold and jewelry involved in his dealings with Richter. What do you make of that? He also said that when the war was over, he fled south with a group including Richter and Werner's father. Do you think they could have had that sort of stuff with them?"

Tev nodded. "I could imagine the three of them hightailing it out of Bergen-Belsen, changing into civilian clothes, and carrying their loot with them. The SS had access to prisoners' jewelry, gold dental fillings and so forth. When they passed by this pond, it must have seemed a natural place to hide it before going on to Frankfurt. They could easily have placed their records and their loot in waterproof bags and dropped them to the bottom. For all we know, Richter and the senior Werner eventually used their loot to found the company. Ironic. But you can't go fishing underwater every time you need records and lists, so Richter may well have decided to build a scenic stone bridge over his pond and use it as sort of an outdoor safe. Employees were kept off it by security guards, but he could come and go as he pleased. That way, if his office or home was raided, there would be nothing found. It's an intriguing possibility. We can't prove it in the dark, but I suspect that when Richter moved the stone out, his actions would be hidden from the plant windows. Now he's scared and he's taken the lists out, and probably all his Libyan material as well. I hope he's not planning to flee the country. We'd better get Johanna. To borrow from your medical parlance, I think we'll have to pay Richter a house call after all."

On Johanna's next circuit, they were in the preassigned place, waiting.

"Any luck?" she asked.

"Yes and no," Tamir said. "We found the hiding spot, but it was empty. Let's go to Richter's home. I'll direct you, Tev. I've been there before."

Finding Richter's sumptuous home was no problem, but unlike the plant, lights were on. Cars were parked in the driveway and two men stood at the front gate smoking cigarettes, their hands stuffed into their coat pockets.

"Richter's men," Tev said. "He must have a private army. I don't think we can force our way through them without reinforcements and I can't get any at this hour. Damn!"

"What about the police! Slitko?" Goodman said.

"Only as a last resort; I want those documents back in Tel Aviv, not getting 'lost' in some old Nazi bureaucrat's office in Bonn. We'll have to try again tomorrow. Enough for tonight."

At the Werner house, they were about to say good night when Johanna found a note from Trudi, who had long since retired.

"You'd better see this," she said, handing it to Tamir.

The Israeli swore again. "Damn!"

"What does it say?" Goodman asked.

Tamir read it aloud: "Dieter Hoffman called, Peter's technician. He found out that Richter was planning to leave tomorrow at 1200 hours for Damascus. He thought you might be interested."

Tamir put down the note. "Yes, we're interested, but damn! Syria! We'll never get to him there. That's not like South America."

Carlights shone through the kitchen window. "Someone's here," Johanna said worriedly.

Tev looked out into the driveway. "Relax. Its Zvi."

246

Zvi's face was flushed. "I thought I'd catch you here. I just talked to Shlomo. There's lots to go over. Can we talk here?"

Tev looked at Johanna. "Trudi and the children are sound sleepers," she said, "just do it softly."

Zvi gave a note to Tamir, who read it while Tev quickly briefed Zvi on their unsuccessful search for the files and the news that Richter was leaving the next day for Syria.

"He'll liquidate Frankfurt Pharmaceutica and have the funds transferred through Swiss accounts to himself and his old cronies," Zvi said. "He'll pension *himself* off now. Once he leaves Germany he'll become a much more difficult fish to catch. And to catch with the goods, I might add. I suppose if Richter hadn't known of Stern's death, he wouldn't have panicked, but this way he's not sure who knows what."

"Isn't there still some way to get the documents?" Johanna asked.

"We can't storm the house," Tev said, "if that's what you mean. I've had second thoughts about that."

"But when he leaves tomorrow," Zvi said, "he'll carry them with him. They're too valuable to leave in his luggage. Luggage can be misplaced, lost. Shlomo thinks that makes the most sense. He's got a hunch that the plans for the plant were given to him so that he can blackmail the Libyans if necessary. Or, to show the Syrians or Iraqis that he can do for them what he did for the Libyans. His going to Damascus fits right into that."

Tev nodded. "We must get the files, but how? I have a feeling Richter's trip to the airport will be a carefully thought out one, difficult to intercept. There must be another way." He pulled out his pipe and clamped it between his teeth while he pondered his next words. "Ah-hah. Yes. There is, but it will be a very

247

dangerous way, of that I am sure. And it will mean Paul and Johanna—and especially Paul—will be exposed to possible death."

"I can't say I'm thrilled about that," Goodman said dryly, "but I'll hear you out."

"I think Shlomo's correct," Tev went on, "I think Richter will carry the papers with him, probably in his briefcase. It just might be possible for you to approach him in the airport, talk with him and at the right moment walk off with the briefcase. You would then transfer it to me and leave the country immediately. Here's my plan . . . " Softly, and with little emotion, he spent the next ten minutes outlining what they would all do the next day.

"It's dangerous," Tamir said when Tev had finished, "but we have no choice. We'll be back at 9:30 tomorrow morning," he said to Goodman and Johanna. "Be packed and ready. Unless, of course, you change your mind."

"We won't change our minds," Johanna said, looking at Goodman.

Zvi smiled. "You've both been very helpful to me, and I appreciate it. I'm especially grateful to you, Paul, for explaining to us the properties of 368 and its poisonous 'cousin.' I think I have a much clearer idea of what has to be done to deactivate its production." He held up the note from Zvi. "Shlomo has obtained enough information now to give the go-ahead for the raid on Rabta, but he's very concerned that someone be there who knows how the chemical warfare agents are put together. So, Paul, I won't be with you tomorrow. My work here is over. Nora, one of Tev's people in Germany, is going to put me on an early morning flight to Israel. Shlomo wants me with the commando unit. I can use my ability to speak Arabic plus my knowledge of the toxic compounds to take a team into the core of the Libyan plant, destroy their setup and get out before the fighter-bombers arrive to demolish the plant. If I leave now, I'll have one or two days to train with them. I'll be very frank with you—my job would be a lot easier if I had the detailed plans of the plant that Richter's probably carrying with him along with his other files. Remember that tomorrow. Without those plans my unit could take heavy casualties."

"I'm still puzzled by one thing," Goodman said. "How will you catch the Libyans by surprise? Or is that something you can't talk about?"

Tamir smiled. "You're better off not knowing."

"You're all brave men," Johanna said, in a very soft voice.

"We're doing what we have to do to protect our country," Tamir said without bravado. "We don't call it the Fatherland, but we will defend it to the death. Especially against Richter's cronies."

"They're left-over scum from that sewer called the Third Reich," Johanna muttered with obvious distaste.

Goodman took Tamir's hand. "Good luck, Dan," he said somberly.

"Thank you, Paul." Their gaze locked for a moment as they shook hands, then Tamir left with the other two Israelis.

When they had gone, Goodman paced the kitchen nervously. "I wish to hell I was as convinced of Tev's scheme as he is."

"I don't think he's convinced of it either, but we can't let them down."

"Just like that, no questions asked?"

"Richter's a murderer, a war criminal. He's also responsible for my brother-in-law's death. He probably gave the order for us to be killed, and he's helping those who are planning to murder thousands of Israelis.

249

You should have other reasons as well. You walked with Becker at Bergen-Belsen, you know those were not just abstract numbers on those markers. Those were the numbers of bodies. Bodies of people, maybe your own distant relatives. If a person like Becker could hear their cries from the grave, why can't you?"

Her face was flushed. Goodman knew she was right—how could he ignore those cries, when she couldn't?

He looked at his watch. "We only have a few hours before dawn, but I'm not very tired. I don't know if I can sleep."

"I want to lie next to you for a while and have you make love to me," Johanna said. "I'm in a funny mood; I feel like being a little sentimental. Indulge me."

In her room, the shade up and the moonlight streaming in, they explored once more the delights of each other's bodies. Of all the times they had been together in the past two weeks, it was this night that Goodman knew he would always remember if he were to survive the next day's adventure. Not just because of that adventure, but because she seemed to be bursting with love this night, insatiable in her demands for his lips, his face, for all parts of him. And he responded in kind. The first light of dawn was creeping across the horizon before they were finally satisfied. She curled in his arms, but neither had any desire to sleep. They watched the sun rise, in all its beauty and majesty, knowing full well that there would be plenty of time to sleep later, whether in a plane, or in a jail, or . . . wherever.

"I want to get you a belated Christmas present," Goodman said, "but it will have to wait until we get to New York. In New York at this time of the year the stores along Fifth Avenue are a fantastic sight, the

250

Christmas trees in the center dividers on Park Avenue are still up and they're covered with white lights, and even the people are friendly. You'll love it. And we'll be safe there."

"It sounds wonderful."

"But we have to get through the business at the airport before we can think about New York."

"We will," she said confidently. "Remember that first night we met? You looked so surprised when I came into the room."

"I wasn't surprised, I was astounded. The combination of beauty and brains was too much for me. I was smitten on the spot."

She laughed. "I was smitten too, that's why you had so little trouble seducing me. And that skiing holiday, how wonderful that was—and before that, dancing at the Nightingale, hearing the music that you like so much. I really felt I was getting to know you then. But I suppose it was when you stayed to help us sort out Peter's death that I realized how much I loved you, the person, not just you, the romantic American." She snuggled closer.

He pulled the sheet over the both of them and held her close. "For five years, I've been trying to find someone like you, without any success at all, and here I stumble across you because of a drug company trip that I really didn't want to take in the first place." He kissed her gently on the forehead.

Frankfurt, January 17

The full light of the new day drenched the room. "Time to go," Johanna said reluctantly. "I'm going to take a bath and pack before breakfast. Telling Trudi that I'm leaving won't be easy."

"The less she knows, the less she'll be questioned about later by the police."

When they came to the breakfast table, it was nearly nine o'clock. The children were in a hurry for their school bus—they seemed to have returned to their usual patterns with the resiliency characteristic of the young. For Trudi, it was not, and would not be, as easy, but Goodman had the feeling that in time she would remarry and the children would once again have a father, hopefully someone as caring as Peter Werner. When the children had gone and the dishes were cleared, Johanna told her sister that she would be going away for awhile.

"When will you come back?" Trudi asked.

"I don't know," Johanna said. "When things quiet down. I can't say any more now. You'll understand later."

Trudi showed no emotion. Whatever doubts she had about Goodman had vanished with his staying to help her, but enthusiasm was still too much to be asked.

"I want to thank you for helping to learn about Peter's death. From what you've told me, the police believe that Stern had a hand in it, but since he is also dead, they will never be sure. It was that drug, that 368, that killed him, I'll always believe that." She looked away tearfully. "But that is what Peter wanted to do. That was his life."

Her tears were interrupted by the ringing of the telephone. "It's for you," Trudi said, handing the receiver to Goodman. It was Ernst Grundig calling from Berlin.

"I remembered your telling me you were staying at the Werner house, so I called hoping you were still there. I heard about Becker's death. I have a feeling I may have been directly responsible by telling you about him. I only knew him from what others said. He was no saint, that I know, but he did almost come over to us once, so I feel he couldn't have been all bad. In any event, I'm feeling contrite enough to supply you with information I left out of our earlier talks. Maybe it will be of some value. Hans Meyer was picked up by the police in Leipzig and he will be transported to Frankfurt within a matter of days. Inspector Slitko, who apparently knows you, is handling these arrangements. Meyer is part of a group of neo-Nazis operating clandestinely on both sides of the old border. Richter and *his* group of Nazis are probably involved as well. While in custody Meyer blurted out some startling information that apparently very few people know about, not even Richter."

"What did he say?"

"Shortly before Peter Werner's death some of the original rats tested with 368 developed rather obvious abdominal cancer. In other words, the drug is a carcinogen and would never have been approved for human use. That must have been devastating news to Peter Werner."

"If he knew of it before he died . . . "

"Meyer said he did."

Grundig was right, Goodman thought, it must have depressed Peter terribly. It meant, at the least, more years of animal testing to disprove, if possible, any cancer links.

"Well, there's no need for you to continue to get the drug, then. It's useless."

Grundig cleared his throat. "I must confess, I did not want it to use on one of our VIPs. In fact, I'm not quite sure what the people who wanted the drug were going to do with it."

"Who wanted it?"

"The KGB."

"You have no idea why? I find that hard to believe."

"Please, I'm embarrassed enough as it is. I do have some speculations which I'll share with you, but I cannot confirm them."

"I'm listening."

"My Russian contacts told me the KGB needed to analyze 368 because it is somehow related to chemical warfare agents the Libyans and Iraqis are developing over Russian objections. The Russians have been trying to warn NATO and Israel about new acts of terrorism by their former client states in the Middle East, actions which could bring on a war no one in the Kremlin wants."

Goodman said nothing. He had no way of knowing whether the information was already known by the Israelis or not.

"I understand your silence," Grundig went on. "I pass this bizarre news on to you only because I feel an obligation to put this affair behind me. I'm truly sorry about Becker." The phone clicked dead.

Just as Goodman was fabricating some explanation to Trudi for the phone call, a car horn sounded outside.

"Friends are taking us to the airport," he explained to Trudi, ignoring her questions about Grundig's call.

"Will they come in for coffee?"

"No, I think we'll just take the bags out and say our good-byes now. You've been more than hospitable

254

to me and I'm forever grateful to you." Neither Goodman nor Johanna wanted Trudi to meet Tev or Zvi. The less she knew, the more honest her denials would be later.

Johanna and Trudi embraced while Goodman brought the suitcases to the car. Tev had his collar pulled up and his cap pulled down and wore sunglasses. Zvi wore a hat to cover his blond hair. They also wanted as little recognition as possible. "Where's Johanna?" Tev asked.

"Saying good-bye to her sister. How was your predawn trip to the airport?" This had been part of Tev's plan.

"Good, very good. I'll tell you about it on the way."

Johanna came out of the house with tears streaming down her face, her hair flashing golden in the bright sunlight. She quickly got in the back seat. "Drive," she said, wiping off her cheeks. "Drive before she comes out to the car."

Tev gunned the motor and the BMW roared off. Trudi sobbed in the doorway, her figure getting smaller as the car raced down the street.

When they had gone for a mile or so, Goodman told Tev to pull over to the curb. He couldn't keep Grundig's news to himself any longer.

"Unbelievable!" Johanna gasped. "If Richter had waited a little longer, there would have been no need to suppress research into the drug. No government agency would ever have approved it, isn't that right?"

"Absolutely," said Goodman, "and Peter would probably be alive today. Unhappy, but alive."

Tev shrugged. "The laws of fate are strange. What matters is that he thought he had a wonder drug and Richter and Stern thought that, too, and they acted accordingly."

"I suppose you're right," Johanna said, "but still . . . " Her voice trailed off and she was lost in her own thoughts while Tev maneuvered the car through Frankfurt's early morning traffic.

He didn't go directly to the airport. Instead, he drove to Richter's house. "One last look," he explained, "just in case his guard is down." But it wasn't. Two men barred the front gate; both were husky and dark-complexioned. Tev drove past the house, then pulled over to the side. He adjusted the rearview mirror until he had the front door of Richter's house in the mirror.

"We'll wait until we're sure they're on their way, then go to the airport. Don't worry. I'll get you there in enough time to pick up your tickets. Someday maybe you'll make another long trip and visit me in Jerusalem."

"Do you miss Jerusalem?" Johanna asked.

"Very much, especially the Old City, the narrow streets with the Byzantine atmosphere. I love to wander about in there whenever I get the chance. The desert I've had enough of. And Tel Aviv is too much like any other big city, but Jerusalem, Jerusalem is different. Jerusalem, the Golden. And for me, the best part is the Old City. Someday, maybe we'll bump into one another there. What do you say, Johanna?"

A far-away look came into her eyes. "Someday," she said, and shot Goodman a quick glance.

Why not, he thought, why not?

"Ah, here they come," Tev broke in. "Good. That gives us plenty of time at the airport."

Richter, surrounded by two other men, all bundled in overcoats, entered a black Mercedes. Satisfied that Richter was actually on his way—the two assistants carried a suitcase and Richter a bulky brief-case—Tev took the shortcut he had mapped out

256

several hours before, one that should get them to the airport a good fifteen minutes ahead of Richter's car. As he drove, he spoke again of what he planned for later than morning.

"Why do you think Richter's bodyguards will let me even talk to him?" Goodman said.

"That's up to you. You've got to appeal to his ego. You admit defeat; you're leaving the country yourself—what better chance for him to rub your face in the dirt."

Can I really do this? Goodman wondered. Stop him as he's walking to his plane?

At the sight of the airport, Goodman felt a cold sweat break out on his palms. It was one thing to talk about these things in the abstract, but quite another to actually take part in them. Tev parked the car in the long-term facilities. "Someone else will be picking this up in a few days," he explained. "We always try to leave it in the same general area." They each carried their bags to the terminal, and at the Air Canada desk, Goodman and Johanna purchased their tickets on the 12:10 flight to Montreal and checked their luggage. Tev and Zvi carried two identical tote bags, one of which they gave to Goodman. It was 10:30. They walked slowly to the international departure area, Tev making conversation about the weather, the stores, anything. But Goodman was growing more and more nervous, his hands now ice cold.

"Stop here," Tev said. He positioned Goodman so he would face the door that Richter would be coming through. "This is where we part, my friend, at least for the moment. Get him to sit with you on one of those benches away from the newsstand. Be persuasive. Titillate him, appeal to his ego, but don't mention Libya. That might provoke him. When you're sure the briefcase has the documents we want, have his goons go for coffee or a paper, then pull your left ear twice. Two minutes later, Zvi will create the diversion there." He pointed to the far end of the terminal. "Two seconds later, the last act begins. The rest you know. Well, what do you think? Crazy, huh? I've seen crazier things than this work, believe me. Okay, I'll leave you two alone now. Shalom." He walked off toward the balcony, sunglasses now in his pocket, collar pulled down, hat brim up. Zvi gave a short wave and walked off in another direction.

"I'm scared to death," Goodman confessed to Johanna. "It's worse than the night at Stern's house. There, everything happened so quickly. I didn't have much time to be afraid."

Johanna squeezed his hand. "You'll be all right, I know you will. I love you."

He looked into her eyes and saw his fear mirrored in them. His stomach was knotting up. But he told himself he would do it. If not for himself, then for her, for Peter Werner, for the dead at Bergen-Belsen, for those who would die in Tel Aviv and Haifa if he failed.

She kissed him on the cheek. "I'll see you later," she whispered.

He watched her walk away, watched her turn once and smile at him, watched her disappear into the recesses of the terminal. She would be waiting at the Air Canada gate that was right across from the Lufthansa gate for Middle Eastern flights. Tev and Zvi had done their planning well, now it was up to him. Planting his feet solidly beneath him, he fastened his attention on the door that Richter should be coming through at any moment. The wall clock read 10:45.

Frankfurt International Airport,
January 17

The minutes ticked away. With his hands clenched tight, knuckles white from the pressure, Goodman waited. The wall clock read 10:46, 10:47, 10:48. Around him the terminal was bustling. Cheerful music played in the background; 10:49, 10:50. People smiled, laughed. Everyone was in a good mood this morning. Even the airport police seemed less austere; 10:51, 10:52. Still no Richter.

At 10:53, the anxiety that had been building up inside Goodman dissipated as Richter walked briskly into the terminal. Goodman shifted into his line of approach and moved slowly toward him and his two bodyguards. Richter saw him and stopped in his tracks, perplexed, then glanced around at his security and waited for Goodman to come closer.

"I can't say I'm delighted to see you here," Goodman said, feigning surprise and indignation. "I'd hoped you'd be in jail by this time."

Richter glared at him. "Taking my advice and going home? No, I'm not in jail and don't intend to be. A long vacation is what I need, especially from the likes of you."

"I may have been a nuisance," Goodman said, "but you seem to have taken it in stride. In a perverse sort of way, I suppose I must acknowledge your talents."

Richter's eyes softened to the barest degree and Goodman knew he must proceed forcefully. It was now or never. "You outwitted everyone, even the police. How did you manage to do it?"

Richter relaxed. "So you want to feel even more unhappy before you leave? That invitation I find hard to pass up. First," he signalled to the thug on his right, "Ali will make sure you're not wearing a tape recorder." The Arab frisked him quickly but thoroughly, and grunted his approval.

"Now. What would you like to know," Richter crowed triumphantly.

"Could we sit down?"

Richter grinned, totally enjoying the situation. "Why not?"

Not believing the ease with which he accomplished it, Goodman led him to the area that Tev had designated. The two goons stood close by.

Richter sprawled comfortably in a plastic chair, his red face still smiling. "You *have* been a nuisance," he admitted, "and I suppose it was time for me to retire anyway." He shrugged. "But such a nuisance, and over nothing."

"What do you mean?"

"Peter Werner was never killed on my orders. Nor on Stern's."

"I find that hard to believe."

Richter shrugged. "My whole preoccupation with this 368 business was to see that the drug was never released on the market. I didn't want the company to have a wonder drug because that would inevitably lead to feature stories on our company by the newspapers. You learned about the experiences of Peter's father and myself during the war; for various reasons, I didn't want that brought to public attention. I decided to suppress all good things about the drug and play up all the bad aspects. Stern helped by falsifying the patient studies."

"How were you going to handle the non-German trials? Beat up the investigators, like you did with Hawthorne?"

260

"That was unfortunate, but he was much too enthusiastic. No, I planned to simply withhold all those funds I so generously offered you. Blame it on the economy. Oh, Peter would have been upset for a while, but a pay raise and extra vacation time would have quieted him down. Remember, I was like an uncle to him; in the end, he would have done as I wished. Even though he proved more troublesome than I had anticipated—doing his own patient studies, for example, I never had any intention to kill him. And I didn't. I didn't tamper with his car or poison him, and neither did Stern."

"Then who killed him?"

"Nobody. It really *was* an accident, at least as far as I know. Whether you believe me or not, is your business."

Goodman was surprised. But why would Richter lie now?

"Peter carried a pistol. He must have been afraid of something."

"He had become somewhat paranoid over this 368 business, so I'm not surprised to hear he carried a gun. What he thought it would protect him from, I have no idea."

"What about the car brakes on Johanna's Porsche? I suppose you didn't have one of your henchmen tamper with them, either."

Richter looked at him disdainfully. "I see you too were bitten by the same bug that Werner was. I ordered no tampering with *any* cars. The tension must be getting to you. It's well that you are going home."

"All right, forget the cars and the pistol, but you won't deny you gave Stern the orders to do us in?"

"No, I won't deny it. In front of the police I couldn't show my puzzlement over the story that Stern and his assistants had shot it out. I had a feeling that you

261

knew more about that than you have let on. And if you knew, the police knew. So you see, by confronting me like that you have hastened my departure. You are a nuisance. What really happened to Stern?"

"The Arabs wanted more money for getting rid of us than he was going to pay them," Goodman lied. "It was as simple as that."

"And the police know this?"

"Yes."

"So I am right in getting out of Germany!"

"Yes, you're very clever. But I still can't believe that Stern didn't figure in Peter Werner's death, whether directly or indirectly."

"In a way, I suppose he did, but indirectly. At their final meeting, Stern told him about his father and myself. That was to show him what publicity about the 368 project could ultimately disclose. It's possible the news proved too much for him to handle later and in his already distraught state, he lost control of his vehicle."

"The publicity would also harm *you*, wouldn't it?"

"Yes, of course."

"Aren't you leaving something out? Wouldn't publicity perhaps even disclose your payments to your old friends? Don't look surprised. Becker told me all about it."

"Payments to old friends? So, Becker told you about that too? What a talker that Becker was. But he doesn't talk much anymore, does he? No, not Becker, thanks to Stern. Stern was good, loyal; there aren't many like him. He was part of our organization because he wanted to be, he was too young during the war to do any soldiering. But his uncle was Otto Strumpf and Otto had taught him well. Frankly, I am surprised that he had a falling out with Mohammad and his brother

262

before they silenced you and your girlfriend, but maybe they did want more money. I hired them because they're reliable and cheap, but who can tell with these people? The two I brought today are even more fanatical, the kind that would shoot up an airport at the drop of a hat,'' he winked mischievously. ''They understand the need for discipline. But that's not what you want to hear about. You want to know about my 'payments to old friends,' don't you? You want to know why a company like mine exists mainly as a means of giving away its hard-earned profits to a bunch of people who don't even live in this country, much less work for the company? That's what you want to know, isn't it?''

''Yes.''

''I have a list of names that I've kept well hidden. Very formal, like a telephone book, 200 to 300 names and addresses. It goes with me now.'' He tapped his briefcase smugly. ''And don't get any wild ideas about grabbing it. Ali and Awad would kill you instantly with their bare hands if necessary.''

He leaned forward and spoke emphatically. ''Yes, my friends are the defenders of their Fatherland in the best sense of the word. What you don't appreciate, what you can never appreciate, is what the names mean to me, because I knew most of the people, saw them doing their duty for their country. They were SS scientists, doctors like myself, orderlies, anyone who could help us. Advancing medical science, that's how I look at it. Of course, you don't agree, so you needn't make faces at me. Yes, I see your anger under the surface, but what's the point in arguing with you? You see, all these people had to be looked after. Those who got away, that is, those who didn't have resources to fall back on, those who couldn't work at new jobs for one reason or another. These people had to be supported. They had

263

earned it by their devotion to their country. I say country because I think you can understand that better than devotion to their Führer. Americans have a blank when it comes to understanding our devotion to Adolph Hitler and as a Jew, you can't even discuss the matter rationally. So I talk of devotion to country. Today, many people won't appreciate my work. That's why I must go away. Even in Germany my efforts may be misunderstood once news leaks out, and I fear with Stern's death and your bothering the police so much, it will only be a matter of time before they make my life here uncomfortable, to say the least. Meanwhile, there are other countries that will appreciate my presence."

"And your company?"

"I'll sell it. For my people, I would do anything. My old camp commander is in Buenos Aires; a great man, though perhaps not by your standards. But they are not just in Argentina, they are everywhere, hundreds of them. Some are in Libya and Syria, which is where I'm going now." He patted his briefcase. "In fact, there are documents in here that people would die for, papers that could plunge the Middle East into even worse chaos than it's already in. And best of all, schemes that will bring money for my pensioners even after the company is sold. But that is something you know nothing about . . . and you won't." He quickly changed the subject, not realizing he had already told Goodman exactly what he had wanted to know. "One of the camp guards at Auschwitz, a real disciplinarian, do you know where he is now? Guess! Guess!"

"I have no idea."

"In New York City! He's living right there with you in New York City literally surrounded by the relatives of his former prisoners. How can this be? That's what you're thinking, isn't it? It's simple, really. After the war, he left the camp—like the rest of us—buried or

264

burned his uniform and mingled with the refugees. From the refugee camps, our comrades went all over the world. Any why not New York? Who would look for an ex-SS man there? There he married and raised a family. But he hurt his back and can't work regularly. Through the contacts we have set up, we learned of him and put him on our pension plan. Three times a year, he gets a check from Uncle Franz in Frankfurt. But, of course, the money goes through Switzerland or Hong Kong to keep the company books clear—unless people begin checking too closely."

"You pay any ex-guard who needs it?"

"No, just the Germans. And only those who are poor. The Ukrainians and Lithuanians and the rest who worked for us are on their own. Some of them are in the United States also. Your country loves anti-Communists, so that's what they call themselves now. But they not Germans, so I don't pay them. I have my principles."

Goodman had to restrain himself from smirking. Principles! Richter talked of principles! "It's a good story," he told the beaming Richter, "so good I'd like to tell it again in the newspapers, on radio, on television."

"But who'd believe you?"

"If I had the documents in your briefcase, people would believe me."

Richter laughed. "Yes, you're quite right." He clutched the briefcase more tightly. "But I'm afraid that's not possible. No, we had to do what we did to protect the company and its sources of income; 368 would have exposed us too much, as you must have suspected. But why are *you* smiling?"

"Now it's my turn to surprise you. You didn't have to do anything. The drug's carcinogenic."

"What? Who told you?"

265

"First, send your goons for coffee or something. They give me the chills."

"Your story better be worth it," Richter said. "They won't be far if I need them. Ali, take Awad for coffee. Go to that stand over there where you can keep an eye on me." The two did as they were instructed. "Now, who told you this?" Richter demanded.

Goodman repeated the lines he had rehearsed carefully the night before, ones that he hoped would interest Richter enough to momentarily drop his guard. "Grundig. The Communists had tested a similar compound and gave up on it for that reason. They knew all about 368 and they couldn't care less."

Richter's face fell. "The damn drug was carcinogenic all the time?" He shook his head in disgust. "So you get the last laugh after all, if one can believe the Communists."

"Grundig is reliable. He told me about Becker and Becker told me the rest. He also seemed to know about Hans Meyer fleeing to East Germany, which really surprised me."

"Not so strange," Richter said with a grim smile. "Some of the most fervent new followers are in Leipzig, in Dresden and in hundreds of towns and villages ruled by the Communists since 1945. They are sick of the left-wing drivel that reduced their part of the Fatherland to a serf-like lapdog of the Soviet Union. Nationalism is surfacing again throughout Central Europe and it will result in a new, strong and united Germany, a Germany that takes orders from no one: not the United States, not NATO, not the Russians. And in the old German Democratic Republic, they are the most restless of all—the political parties that campaign on these themes will do well. Wait till times get bad and jobs are scarce. Our appeal will be to strengthen the Fatherland, the new United Germany. And here, too," he said smiling, "our parties will do well, even in socialist Frankfurt. Did you know that after the World Cup victory 20,000 people wanted to appoint the winning goal scorer the 'Führer?' Yes, they said it loud and clear, they didn't care who heard them. That's when I knew our ideas could once more take root. Believe me, one of these days a new leader will come forward to restore German greatness. That leader will be a product of our movement. Then we old men will once more be of use to our Fatherland."

"Wait a minute," Goodman said patiently. "You said at the dinner at the Red Lion that the neo-Nazis and the anarchists were all the same, and you had no use for either. You said that even before Eriksson spotted Hans Meyer in the crowd."

"Of course, I said that, you fool. What did you want me to say? That I secretly longed to be down there with them? But Hans Meyer, that is another story. He was a valuable member of a certain work section at my plant. I should not have exposed him to any undue attention and, in retrospect, it probably was a mistake to let him join our new party, but he was so enthusiastic. How could I resist? When Eriksson saw him in the street, I knew there would be trouble and away he went. So Grundig knew about him? That can only mean the hardliners also knew about our dealing with some of the Middle Eastern countries, business you would love to learn about, I'm sure. The North Koreans were probably jealous and wanted to do the job themselves. They always need hard currency. That's why I wonder about Grundig, but it doesn't matter now. Once we restore the Fatherland, then perhaps we'll settle those accounts." His eyes took on a faraway look. "You know," he mused, "it's a pity that you, as a cardiologist, can't appreciate the experiments I did during the war. It's all in my briefcase, you know. And some of the things we did then we are doing now, yes, even now, but in more sophisticated ways. Perhaps soon you'll know it, you and all the other Jews. It will be 'published' someday.'' He smiled when he said 'published,' then grew serious again. ''You Jews blame us for everything when it was actually your own fault. You know it very well, you've provoked the German people for centuries and now you're doing the same with the Arabs.'' He went on and on, his face growing more livid. He really believes it, Goodman thought, which makes it easy to hate him, easy to see him killed. Johanna was right. Why waste sympathy on this miserable creature who probably ordered scores of people killed in the camps. He pulled his left ear twice, the signal to Tev, relieved that now he could do it without any moral qualms, then sat and listened to more of Richter's drivel as he waited for Zvi's diversionary device to be detonated. It went off with a huge roar.

The eyes of all the passengers in the waiting area fastened on the mezzanine shop spewing forth thick black smoke. The illusion that a bomb had gone off was complete. Within seconds the screams and shouts began. Confusion reigned. Goodman waited breathlessly for evidence that Tev's 'plastic' pistol— with its silencer—had played its part.

As if on cue, Richter's head snapped up and he slumped forward, his eyes rolling upward. Tev's shot from the shopping arcade above him could not have been more accurate. The two goons at the coffee stand were still transfixed by the commotion at the other end of the terminal. Unnoticed in the chaos, Goodman propped Richter in the chair, threw the briefcase into his tote bag, and walked quickly toward the Air Canada gate. It was eleven thirty. It had been easy so far. He wondered if Richter was still propped up, but he dared not look back. As he walked on, security guards ran past him toward the smoke, their weapons in their hands. The shouts and screams went on unabated, clusters of frightened passengers running in every direction. Goodman picked up his pace, not wanting to draw undue attention by too much nonchalance. In a minute, he was out of the main passenger area and into the Air Canada boarding zone, where there was relative calm.

Ahead of him, Johanna waited anxiously, sighing audibly with relief as she saw him striding toward her. From a side door, Tev entered the boarding area and stood by Johanna, one hand in his pocket, the other holding his own tote bag. It all seemed dreamlike. Goodman approached Tev, put down his bag, nodded a greeting. "Dan's stuff is in here, too," he whispered triumphantly. Tev's face lit up with relief, then he picked up Goodman's bag and retreated toward the side door, his eyes still searching the corridor behind Goodman for a sign of trouble.

"Please continue boarding," the ground attendant urged the Air Canada passengers, "we plan to depart for Montreal on time." Goodman took Johanna's arm and joined the end of the line of waiting passengers. The wait seemed interminable as the line inched forward. Meanwhile, the commotion around the explosive device had abated; the police were urging calm.

"Did it go all right?" Johanna whispered.

"Perfectly. Richter couldn't miss an opportunity to do some bragging."

"He admitted everything?" she said, wide-eyed.

"He confirmed everything that Becker and Tev had told us."

"What about Peter? How did he arrange the accident?"

269

Goodman paused. "He disclaimed any involvement in it and strange as it sounds, I believe him. Remember, Stern denied it too."

"Then how?"

"Apparently, during the last meeting they had together, Stern told Peter about his father's involvement with Richter in the SS and how publicity about 368 would bring all that into the open. Richter thinks, and I suspect he's right, that Peter became so distraught with that news, plus what he had heard about the compound's cancer causing potential, that he let himself get distracted while driving and made a fatal mistake."

"Then his death was just what it appeared to be, an accident! All of this trouble because we investigated an accident that really was an accident." She shook her head in disbelief.

"But we've done some good after all, even if we did blunder into things for the wrong reasons. That's some consolation." He stopped abruptly. From the main lounge had come one of Richter's bodyguards. He was in full flight toward the Air Canada desk, his eyes fixed on Goodman.

"Hey there, you! Stop!" A security guard shouted in vain as the huge hulk shoved him aside and kept coming down the corridor. The Libyan, Ali, veered wildly, first to the left and then to the right, but he kept his eyes focused on Goodman and his tote bag. Shoving another security guard aside, Ali came straight at Goodman, his face a mixture of anger and hate. Goodman pushed Johanna to the floor while the other passengers on the line fled. But the Libyan never reached them. The security guard tackled him from behind, then wrestled him to the ground. Before he could subdue him and before the other guards could assist, Ali had wrenched away the guard's pistol. He aimed it at

Goodman and began firing. Goodman felt a sharp sting in his shoulder and fell against the gate.

Johanna rushed to him. "Paul!"

"Stay back!" Goodman shouted, "I'm all right." But she kept coming. He saw Ali raise the pistol again. He knew what would happen. The bullet meant for him would hit her instead. She would die in his arms and with her would die the love she had brought to his life. He would have to face Slitko again, answer those questions the police would relentlessly press on him, questions that he would not know how to answer to protect Tev and what was in the briefcase.

He saw all this flash before his eyes as Johanna came toward him, as in some terrible slow motion fantasy, while the Arab aimed the pistol. But before he could fire, blood poured out of the Arab's face and neck; submachine gunfire racked his body as the alerted security police methodically cut him to shreds.

"Paul, are you all right?" She hovered over him, confused and frightened, tears forming in her eyes.

He touched his shoulder. It smarted, but he was able to move it easily. "Nothing broken. I'm fine. It must be only a flesh wound. Don't draw any attention to it, or the airlines people or the police will insist on my getting medical attention and we'll miss the flight. We don't want to be around for any investigations."

A crowd of police and passengers had gathered around the body of the dead Libyan. Other passengers still lay sprawled on the floor, afraid to get up lest the shooting resume. The police signaled to the airlines desk to get the passengers on the plane.

"Continue boarding, please," the loudspeaker announced. "We want to clear this area immediately. Continue boarding."

Passengers slowly got to their feet, talking excitedly to one another, pointing in the direction of the policemen going through the pockets of the dead gunman. Goodman rose cautiously, leaning on Johanna for support, trying not to bring attention to his wound.

The terminal was now a hubbub of noise. Police were everywhere, patroling in twos and threes, intently watching the crowds of passengers and their relatives and friends. Over the noise of the terminal came the distinct wail of sirens, as yet more police arrived.

The boarding continued at a snail's pace, the airline personnel still too shaken to perform with their usual proficiency. Shouts and screams came from the main lounge area and then more shots, followed by more shouts and screams. The police in the Air Canada Terminal drew their guns and ran toward the main lounge.

"I hope that isn't Tev or Zvi," Johanna said.

"More likely it's the other Libyan," Goodman responded, but he too was concerned that the Israelis might not have made it to their flight.

The police now began scrutinizing all the passengers more carefully. They separated several swarthy young men from the others and led them to a separate area, apparently for questioning.

They're looking for Arabs, no doubt about it, Goodman thought; a good sign. The boarding line was moving more quickly now and it would only be a few minutes before their turn would come. A new group of policemen now appeared in the terminal. Goodman recognized one of the faces immediately, although the man wore no uniform.

"It's Slitko!" Goodman said, nudging Johanna. "How does my shoulder look?"

"There's some dried blood around a cut in the garment. It's not that noticeable, but if he looks closely, he can't miss it. I don't see what we can do to make it look any better."

"We'll just have to take our chances."

Slitko saw them as soon as he scanned the line of passengers. He stopped abruptly and did a double-take. A puzzled look on his features, he approached them warily.

"I didn't expect to find you two here. When did you decide to leave?"

"Last night," Goodman answered promptly. "As you told us, there was no point in staying any longer."

"And Miss Bauer?"

"She's decided to return my visit and spend some time in Montreal and New York."

Slitko saw the rip in Goodman's coat. "What happened there?"

"Ah, just a scratch during the commotion."

Slitko's eyes met his and locked in. Goodman stared back. He could almost feel the policeman's gaze pierce his veneer of confidence, but Slitko's face betrayed no emotion.

"Yes, quite a commotion," Slitko said, at last. "It might interest you to know there was another victim beside that fellow. Two more to be exact, one a German."

"Oh?"

"Franz Richter was shot in the back of the head in the main lounge. No witnesses, nobody noticed anything. Strange, isn't it? Imagine getting a pistol into such a secure area." He continued to stare at Goodman. "It's almost as much a mystery as trying to figure out where the bullets that killed Stern's intruders came from."

"Richter!" Johanna interrupted quickly. "I can't say I'm sorry. What a nice going-away gift." Her bravado gave Goodman a boost. Better that than panic.

Slitko stared at the two of them very carefully, saying nothing. He spoke only when one of the airline attendants came up to him. "These are the last passengers to board. Unless you're going to detain them, they better get on."

"Detain them?" Slitko said without emotion, "and ruin their trip? I don't think so. This doesn't concern them." He looked back at Goodman. "An unfortunate coincidence that you two should be caught up in this. Have a good flight, and when you arrive home, Dr. Goodman, visit your tailor and have that garment fixed." He turned and strode off, his assistants at his heels.

"Nice chap," Goodman said with relief, "nice chap. There won't be much of a fuss about this at all. Slitko's as glad as we are that Richter's dead. He doesn't care how it happened. You'll probably be safe now, if you wanted to stay. But if you still want to go . . . "

"Of course I still want to go! Sometimes I don't think you're as smart as you say you are."

"Great! Let's get the hell out of here. I came to Frankfurt for a three-day drug conference and I wind up spending two weeks falling in love and dodging people who want to kill me. I think I've overstayed my welcome."

He put his arm around Johanna and walked with her down the boarding ramp. Through the windows of the ramp they could see snow beginning to fall. "That's good snow," Johanna said smiling, "the kind we had in Herzberg, not the kind we had in Kaltenweide."

Good snow, thought Goodman, remembering their hours of bliss in the Harz, trying to blot out all the rest: Peter Werner's death, Becker's confession, the meeting with Grundig, the night at Stern's house, today's bloody scene at the airport. Blotting it all out, except perhaps for a vision of Tev and Zvi flying high above the snow to Tel Aviv, the tote bag and its contents

274

safely in their hands, Tev's special 'plastic' pistol—now in ten disconnected and unrecognizable pieces—in his camera kit.

"What are you thinking about?" Johanna asked as the plane slowly taxied into position on the runway.

"I was thinking of Tev and Zvi. They did a damn good job. They should feel proud of themselves."

"You didn't do so badly yourself," she said softly.

He took her hand in his. She was right, he decided with a sense of satisfaction. He hadn't done badly at all.

"I wonder what's in the tote bag they gave you. It didn't feel like it was empty."

He groped around at his feet for the tote bag. "Nothing but folded newspapers," he answered as he rummaged through the bag's interior. "Wait a minute. Here's something else." He extracted a thick paperback book and showed Johanna the front cover. *Israel on Ten Marks a Day*. "Looks like Tev did some shopping at the airport newsstand."

Johanna grinned. "Think he's giving me a hint?"

"I think he's giving us *both* a hint," Goodman said, adding his laughter to hers.

Epilogue

Israelis Raid Libyan Chemical Plant

Jerusalem, January 19 (AP). The Israeli Government today announced that a chemical weapons plant 60 miles southwest of Tripoli had been destroyed in military action yesterday night. According to the Israelis, the raid, since denounced by the Libyan Government as an "unconscionable violation of national sovereignty," was carried out by a special commando unit of the Israeli Defense Forces and a squadron of fighter-bombers equipped with laser-guided air-to-ground missiles. In its announcement, no mention was made of how the commando unit reached the plant, nearly 1000 miles from Israel, nor of Libyan casualties. Israeli casualties were light; five commandos wounded, none seriously. All planes returned safely.